Key Issues in Environmental Change

Series Editors:

Co-ordinating Editor

John A. Matthews

Department of Geography, University of Wales, Swansea, UK

Editors

Ray S. Bradley

Department of Geosciences, University of Massachusetts, Amherst, USA

Neil Roberts

Department of Geography, University of Plymouth, UK

Martin A. J. Williams

Mawson Graduate Centre for Environmental Studies, University of Adelaide, Australia

Preface to the series

The study of environmental change is a major growth area of interdisciplinary science. Indeed, the intensity of current scientific activity in the field of environmental change may be viewed as the emergence of a new area of 'big science' alongside such recognized fields as nuclear physics, astronomy and biotechnology. The science of environmental change is fundamental science on a grand scale: rather different from nuclear physics but nevertheless no less important as a field of knowledge, and probably of more significance in terms of the continuing success of human societies in their occupation of the Earth's surface.

The need to establish the pattern and causes of recent climatic changes, to which human activities have contributed, is the main force behind the increasing scientific interest in environmental change. Only during the past few decades have the scale, intensity and permanence of human impacts on the environment been recognized and begun to be understood. A mere 5000 years ago, in the mid-Holocene, non-local human impacts were more or less negligible even on vegetation and soils. Today, however, pollutants have been detected in the Earth's most remote regions, and environmental processes, including those of the atmosphere and oceans, are being affected at a global scale.

Natural environmental change has, however, occurred throughout Earth's history. Large-scale natural events as abrupt as those associated with human environmental impacts are known to have occurred in the past. The future course of natural environmental change may in some cases exacerbate human-induced change; in other cases, such changes may neutralize the human effects. It is essential, therefore, to view current and future environmental changes, like global warming, in the context of the broader perspective of the past. This linking theme provides the distinctive focus of the series and is mentioned explicitly in many of the titles listed opposite.

It is intended that each book in the series will be an authoritative, scholarly and accessible synthesis that will become known for advancing the conceptual framework of studies in environmental change. In particular we hope that each book will inform advanced undergraduates and be an inspiration to young research workers. To this end, all the invited authors are experts in their respective fields and are active at the research frontier. They are, moreover, broadly representative of the interdisciplinary and international nature of environmental change research today. Thus, the series as a whole aims to cover all the themes normally considered as key issues in environmental change even though individual books may take a particular viewpoint or approach.

John A. Matthews (Co-ordinating Editor)

Forthcoming titles in the series

Atmospheric Pollution: an Environmental Change Perspective (Sarah Metcalfe, Edinburgh University, Scotland)

Biodiversity: an Environmental Change Perspective (Peter Gel, Adelaide University, Australia)

Climatic Change: a Palaeoenvironmental Perspective (Cari Mock, University of South Carolina, USA)

Cultural Landscapes and Environmental Change (Lesley Head, Wollongong University, Australia)

Environmental Change at High Latitudes: a Palaeoecological Perspective (Atte Korhola and Reinhard Pienitz, Helsinki University, Finland & Laval University, Québec, Canada)

Environmental Change in Drylands (David Thomas, Sheffield University, UK)

Glaciers and Environmental Change (Atle Nesje and Svein Olaf Dahl, Bergen University, Norway)

Natural Hazards and Environmental Change (W.J. McGuire, C.R.J. Kilburn and M.A. Saunders, University College London, UK)

Pollution of Lakes and Rivers: a Palaeoecological Perspective (John Smol, Kingston University, Canada)

The Oceans and Environmental Change (Alastair Dawson, Coventry University, UK)

Wetlands and Environmental Change (Paul Glaser, Minnesota University, USA)

Environmental Change in Mountains and Uplands

Martin Beniston

Professor of Geography, University of Fribourg, Switzerland

A member of the Hodder Headline Group
LONDON
Co-published in the United States of America by
Oxford University Press Inc., New York

First published in Great Britain in 2000 by
Arnold, a member of the Hodder Headline Group,
338 Euston Road, London NWI 3BH

http://www.arnoldpublishers.com

Co-published in the United States of America by
Oxford University Press Inc.,
198 Madison Avenue, New York, NY10016

British Library Cataloguing in Publication Data
A catalogue record for this book is available from the British Library

Library of Congress Cataloging-in-Publication Data
A catalog record for this book is available from the Library of Congress

ISBN 0 340 70638 4 (hb)
ISBN 0 340 70636 8 (pb)

1 2 3 4 5 6 7 8 9 10

Production Editor: Wendy Rooke
Production Controller: Fiona Byrne
Cover Design: Mouse Mat Design

Typeset in 10/11½ Palatino by Academic and Technical, Bristol
Printed and bound in Malta by Gutenberg Press

What do you think about this book? Or any other Arnold title?
Please send your comments to feedback.arnold@hodder.co.uk

Contents

Figures

Tables

Boxes

Acknowledgements

In the 18 months which this book required for completion, I received help from a number of my graduate students at the Department of Geography of the University of Fribourg, Switzerland. Their input was particularly valuable in terms of checking through bibliographic references, searching for information, and preparing summaries of the most up-to-date research results in the various fields discussed in this book. I would therefore like to express my gratitude to the following persons (in alphabetical order): Grégoire Bourban, Jean-Michel Gardaz, Franziska Keller, Pauline McNamara, Frédéric Roux, Marc Valloton, Beat von Daeniken, and Jolanda Würgler. My thanks also to Michèle Kaennel Dobbertin, from the Swiss Federal Institute for Forest, Snow and Landscape Research (WSL) in Birmensdorf, Switzerland, for valuable references for forest and ecosystem impacts.

I would also like to acknowledge the help of a number of colleagues who read through different chapters of the draft manuscript and who made constructive comments on its contents and suggestions for improving the text. These colleagues include Professor Barbara Allen-Diaz (Department of Environmental Science, University of California at Berkeley, USA), Dr Henry F. Diaz (NOAA Environmental Research Laboratories, Boulder, Colorado, USA), Professor Malcolm Hughes (Laboratory for Tree-Ring Research, University of Arizona, Tucson, USA), Professor Bruno Messerli (Department of Geography, University of Bern, Switzerland), and Professor Michel M. Verstraete (Space Applications Institute, Joint Research Center of the European Union, Ispra, Italy).

Martin Beniston
Professor
University of Fribourg, Switzerland
June 1999

Mountains and uplands: an introduction

1.0 Chapter summary

This introductory chapter provides a general overview of the different mountain regions of the world, relevant socio-economic and environmental information, and a brief insight into the causes of anthropogenic perturbations to the natural environment. The issues of global environmental change, which will be addressed in subsequent chapters, are discussed from the viewpoint of economic and demographic factors. A number of definitions are given which provide a helpful reference for the more specialized sections of the book. Table 1.1 and Box 1.1 give comparative information on socio-economic and environmental aspects of countries where mountains are a significant part of their environments. A more generic set of statistics is listed in Table 1.2 for selected countries, in order to emphasize regional differences and sensitivities to the environment and to resource use. A summary is given of the reasons why mountains are important to humankind, beyond the general perception of their marginality and geographical isolation.

1.1 Mountain regions of the world

Mountains and uplands are often considered to comprise some of the world's most extreme environments. However, they are of immense value to humankind as sources of food, fibre,

minerals and water, and they are rich in a variety of living natural resources. Though mountains stand above their surroundings, usually more densely populated plains, they are linked to them in numerous ways, economically, socially and ecologically. Mountains are often perceived to be austere, isolated and inhospitable; in reality they are fragile regions whose welfare is closely related to that of the neighbouring lowlands.

Mountains and uplands may be defined as features of the Earth's surface in which the terrain projects conspicuously above its surroundings, and where the slope of the land distinguishes it from the generally flat plains. While there is no universal geographical threshold above which a topographic feature may be formally defined as a mountain, certain criteria have been established in different countries. For example, in Switzerland, any surface which lies higher than 700 m above sea level qualifies as a mountain zone; this particular level was chosen as a criterion for the allocation of agricultural subsidies to mountain farmers.

Mountains are different from plateaux because of their limited summit area; they may be distinguished from hills by their generally higher elevation. Mountains are normally found in groups or ranges consisting of peaks, ridges and valleys. With the exception of some isolated mountain peaks, the smallest geographical unit for an orographic feature is the range, comprising either a single complex ridge or a series of ridges. Several ranges which occur in a parallel alignment or in a chain-like cluster are known as a mountain system. Systems which are lie across significant latitude or longitude bands form a mountain chain; in some cases, an extensive complex of ranges, systems and chains is referred to as a belt, or cordillera.

Volcanism is one driving force behind mountain building; frequently associated with tectonic activity, volcanoes ideally have a characteristic conical shape composed of lava and volcanic debris, though erosion processes often transform their morphology. Examples in the world include Mt Saint Helens in the United States, Mt Etna in Sicily, Mt Irazu in Costa Rica and Mt Fujiyama in Japan. Numerous volcanoes around the globe are still active and others, while quiescent, are not extinct. Shield volcanoes are isolated systems far removed from plate boundaries, located above so-called 'hot spots' in the mantle underlying the lithosphere; examples of shield volcanoes include the Hawaiian Islands. On a purely local and arbitrary basis, when measured from its base on the ocean floor, Mauna Kea on the Big Island of Hawaii is the tallest mountain on earth (10,200 m).

Plate tectonics are, however, the principal factor for orogenesis; the compression forces associated with convergent (colliding) continental plates are capable of uplift and deformation of crustal strata. This has resulted in many of the major mountain regions of the world, in particular the Himalayas, the Andes and Rocky Mountains, and the Alps, all located in the vicinity of convergent plate boundaries. Mountain chains related to plate tectonics feature the highest summits in each continent: Mt Everest (8848 m) in the Himalayas, Aconcagua (6959 m) in the Andes, Mt Denali (6194 m) in Alaska, Mont Blanc (4807 m) in the Alps are but a few of the most famous examples. Tectonic forces operate on time scales which range from several tens of years to several hundred million years. It is important to differentiate these typical scales from those of erosion processes discussed below, which act on time scales which are between 1

and 2 orders of magnitude smaller. Hence mountains have not only a spatial component inherent to their description, but also a set of temporal scales which govern their cycle from their formation to their ultimate disappearance.

Erosion processes represent a factor other than tectonics which is responsible for orographic features. The Earth's surface is constantly exposed to wind, ice, water, biological and chemical erosion. Rocks of varying composition resist these various erosion agents differently, so that regions of relatively hard rock may ultimately be left standing high above areas of softer, more easily eroded rock. Portions of the intermontane plateaux of the western USA, between the Rockies and the coastal ranges of the Pacific, exhibit examples of such erosion, such as Bryce Canyon in the state of Utah.

1.2 Importance of mountain regions to humankind

Mountain regions have often been perceived in the past – and to some extent still are perceived today in certain parts of the world – as hostile and economically non-viable regions. In the latter part of the twentieth century they attracted major economic investments for tourism, hydro-power and communication routes. Mountains have often been perceived as obstacles to be conquered, but their importance as a major resource for mankind is generally underestimated. Mountains in fact provide direct life support for close to 10 per cent of the world's population, and indirectly to over half (Ives, 1992), principally because they are the source region for many of the world's major river systems. They also sustain many important economic activities, such as mining, forestry, agriculture and energy resources.

Water resources for populated lowland regions are highly influenced by mountain climates and vegetation; snow feeds into the hydrological basins and acts as a control on the timing of water runoff in the spring and summer months. Forests delay the period of snow-melt and extend the period in which water is available for river flow; in addition, forests enhance the infiltration capacity of soils. Regions such as South and East Asia, where almost half the world's population resides, depend largely upon water originating in the Himalaya–Karakorum–Pamir-Tibet region for economic activities such as agriculture, industry and energy production. Changes to the mountain environment, such as shifts in precipitation regimes in a changing climate, reduced snow and ice resources which today feed the major rivers flowing from the Himalayan chain, or deforestation, could significantly alter the flow patterns in rivers such as the Ganges, the Irrawady, the Salween or the Yangtze, and thereby perturb patterns of water use and water management in India, China, Bangladesh, Thailand, Laos, Vietnam, Cambodia and Myanmar.

Biological diversity (biodiversity) is an important issue in mountain regions. Biodiversity is high because of the fact that vegetation patterns are largely governed by climatic factors, which change rapidly with altitude. Over relatively short horizontal distances in high mountain ranges, it is possible to span a wide range of ecosystems, which would otherwise occur over widely separated latitudinal belts. For example, in tropical regions such as the Peruvian Andes, vegetation will go from tropical species at low elevations to tundra and cold-vegetation species in the upper reaches of the Cordilleras. Because of their relative isolation, mountains often harbour unique species of endemic plants and animals, such as the cloud forests in Costa Rica or Papua New Guinea and tropical montane rainforests in many of the equatorial mountain regions of the world. In the eastern Borneo state of Sabah in Malaysia, there are an estimated 4500 species of plants on Mount Kinabalu, which represents 25 per cent of the entire plant species of the United States (Stone, 1992). In terms of fauna, there are more bird species in Costa Rica (850) than in the whole of North America (Boza, 1992), including the legendary quetzal found in the mountains of Central America from southern Mexico to northern Panama. Some of our most common

foodstuffs originated in mountains, such as potatoes (Andes), coffee (Ethiopia and Kenya) and wild corn (Mexico); the genetic reserves found in these regions therefore need to be preserved for future use.

Mountains provide recreational, research and educational opportunities; in terms of scenery, they are often recognized as being among the most spectacular features of our planet. Tourism is now one of the major income sources for many mountain economies; this began as early as the mid-1800s when wealthy British aristocrats began spending vacations in the mountains; mountain climbing was also at that time a key to generating interest in the Alps as a source of recreation. Many fashionable mountain resorts in the Alps, which began centuries ago as poor agricultural communities, today attract numerous tourists: Zermatt, Gstaad, Wengen and St Moritz in Switzerland; Chamonix in France; Berchtesgaden and Garmisch-Partenkirchen in Germany and Kitzbühel, Innsbruck and St Anton in Austria. The attractivity of remote mountain regions, for mountaineering, hiking or even skiing, has increased substantially over the past decade as the costs of air travel to most parts of the world have become more accessible to a larger number of people.

Mountains not only are passive elements of terrestrial environments, but also play a key role in global systems such as climate. They are one of the trigger mechanisms of cyclogenesis in mid latitudes, through their perturbations of large-scale atmospheric flow patterns. The effects of large-scale orography on the atmospheric circulation and climate in general have been the focus of numerous investigations, such as those of Bolin (1950), Kutzbach (1967), Manabe and Terpstra (1974), Smith (1979), Held (1983), Jacqmin and Lindzen (1985), Nigam et al. (1988), Broccoli and Manabe (1992) and others. One general conclusion from these comprehensive studies is that orography, in addition to thermal land–sea contrasts, is the main shaping factor of the stationary planetary waves of the winter troposphere in particular. The seasonal blocking episodes experienced in many regions of the world, with large associated anomalies in temperature and precipitation, are closely linked to the presence of mountains.

1.3 Current environmental and socio-economic information and statistics

Mountains are to be found on all continents, and account for roughly 20 per cent of the world's terrestrial surface area. Figure 1.1 illustrates their global distribution and highlights some of the more important mountain chains.

The proportion of land surface covered by mountains varies from one continent to another; the political boundaries and size of some countries imply that they may lie almost entirely within the mountain realm, such as Bhutan, while others may have no significant orography, such as Denmark. Because of the range and diversity of mountain systems, intercomparisons of environmental characteristics are not simple; in addition, socio-economic systems of particular mountain regions tend to reflect upon the general economic level in which they are embedded; the relative isolation and the harsher environmental conditions of mountains in many countries imply that economic conditions may be less favourable than the lowland regions of that country. Similar mountain regions in terms of their geographic characteristics, for

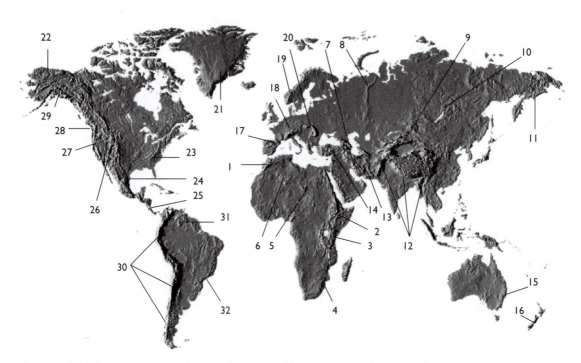

FIGURE 1.1 Major mountain regions of the world. Numbering is clockwise for each continent. Antarctica is not shown; this continent has a number of mountain ranges, including the Vinsson Massif.
Source: Base map © by Digital Widom Inc., USA
Africa: 1. Atlas. 2. Ethiopian Highlands. 3. Kenyan, Ugandan and Tanzanian Highlands. 4. Drakensberg. 5. Tibesti. 6. Ahaggar. *Asia*: 7. Caucasus. 8. Urals. 9. Tien Shan. 10. Altaïr. 11. Eastern Siberian Ranges. 12. Hindu Kush–Himalaya–Pamir–Tibet. 13. Zagros. 14. Anatolian Plateau. *Australasia*: 15. Great Dividing Range. 16. New Zealand Alps. *Europe*: 17. Pyrenees. 18. Alps. 19. Scandes. 20. Carpathians. 21. Greenland Icecap. *North America*: 22. Brooks Range. 23. Appalachians. 24. Sierra Madre. 25. Central American Cordilleras. 26. Rocky Mountains. 27. Sierra Nevada. 28. Cascade Range. 29. Alaska Range. *South America*: 30. Andes. 31. Pakoraimi Mountains. 32. Sierra da Martiquera.

example, may be totally different in terms of their societal or economic conditions.

Table 1.1 illustrates this point by comparing extremes of socio-economic statistics for two land-locked mountain nations, namely Switzerland in the industrialized world and Bhutan in the developing world. Bhutan exhibits many characteristics typical of the developing world: relatively low life expectancy but high population growth rates, high child mortality, low literacy rates, low energy consumption and limited access to consumer goods, and a very high proportion of persons employed in the primary sector (over 90 per cent). Switzerland, as one of the most affluent countries in the world, shows almost the opposite trends in population growth, literacy and mortality rates, as well as consumer and energy-use statistics. Switzerland's low birth rate is insufficient even to maintain its indigenous population at current levels, and the low increase in population over the past two decades is attributed solely to immigration, mostly from neighbouring countries of the European Union.

Table 1.2 provides an overview of population, economic and environmental statistics for selected countries where mountains represent a significant geographical area (IPCC, 1998); the figures in this table are intended to emphasize the points discussed in the context of Table 1.1, namely the cleavage between the industrialized world and the developing world. For several categories of data, in particular urban population, gross domestic product (GDP) per capita and by sector, energy consumption per capita and the proportion of land devoted to pasture, the extremes often illustrate a 'North–South' disparity. North America, Europe and usually Oceania constitute the 'North', whereas Africa consistently represents the 'South'. Asia can more often be grouped with Africa, whereas Latin America's affiliation tends to vary. The division does not fall strictly around the equator, but rather around a latitude band close to $30° N$. The discrepancies are also less accurately described as between industrialized and non-industrialized countries; the striking differences are seen not in the differing degrees of industrialization, but rather of 'tertiarization', the 'North' having service-dominated economies.

A high GDP per capita is associated with a major proportion of service-sector activities, high energy consumption, high urban population and a limited primary sector. It would be possible to conclude that more wealth, as measured by GDP, is generated by the tertiary sector (which may be concentrated in urban areas), and that this wealth is associated with an inefficient use of energy resources. Conversely, economies that are heavily agricultural produce less wealth (in monetary terms). Other important points represented by the data in Table 1.2 include Asia's population density, Latin America and Europe's elevated proportion of protected areas, the high biodiversity of tropical countries and the relatively low biodiversity in Europe.

Several relationships between variables are expressed through rank correlations. Lower water resources per capita is associated with higher density and population, revealing a critical factor in highly populated countries. A higher surface area devoted to pasture lands can be linked to reduced areas of forests, opening up the question of whether forest lands are cleared for grazing. Africa's high fraction of pasture lands compared to Europe, Asia and North America represents a typical situation whereby poorer regions devote more time and energy to agricultural practices, which require less intense financial and technological inputs.

Though the strongest correlations are to be found with GDP data, implying perhaps what conditions may be necessary for 'development', it is imperative to consider that this type of statistic is in no way indicative of living standards and quality of life. For example, it does not provide information concerning health, safety, pollution, poverty or stress among others.

1.4 Environmental stresses: the emergence of the human factor

Mountain regions are particularly sensitive to a wide range of environmental stresses acting at

TABLE 1.1 Comparison of environmental and socio-economic statistics for two landlocked countries of similar size: Bhutan and Switzerland

	Bhutan	Switzerland
Geography		
Location	Southern Asia	Central Europe
Coordinates	27°30′N, 90°30′E	47°00′N, 8°00′E
Area	47,000 km²	41,290 km²
Land boundaries	1075 km	1852 km
Border countries	China 470 km, India 605 km	Austria 164 km, France 573 km, Italy 740 km, Germany 334 km, Liechtenstein 41 km
Population (1996 estimates)		
Population	1,822,625	7,207,060
Population density	38 persons/km²	174 persons/km²
Age structure	0–14 years: 40 per cent; 15–64 years: 56 per cent; 65 years and over: 4 per cent	0–14 years: 17 per cent; 15–64 years: 68 per cent; 65 years and over: 15 per cent
Population growth rate	2.32 per cent	0.59 per cent (1996 est.)
Birth rate	38.48 births/1000 population	11.35 births/1000 population
Death rate	15.28 deaths/1000 population	9.64 deaths/1000 population
Net migration rate	0 migrant(s)/1000 population	4.2 migrant(s)/1000 population
Infant mortality rate	116.3 deaths/1000 live births	5.4 deaths/1000 live births
Life expectancy at birth (total population)	51.46 years	77.70 years
Male life expectancy	51.96 years	74.58 years
Female life expectancy	50.93 years	80.82 years
Fertility rate	5.33 children born/woman	1.47 children born/woman
Literacy rate	42.2 per cent (male: 56.2 per cent; female: 28.1 per cent)	99 per cent (male: 99 per cent; female: 99 per cent)
Economics		
GDP per capita (1995)	$730	$22,400
Labour force by sector	Agriculture 93 per cent, services 5 per cent, industry and commerce 2 per cent	Services 50 per cent, industry and crafts 36 per cent, government 11 per cent, agriculture and forestry 4 per cent
Energy		
Electricity production	1.7 billion kWh	58 billion kWh
Electricity consumption per capita (1993)	143 kWh	6699 kWh
Energy consumption per capita (1994)	15 kg oil equivalent	3694 kg oil equivalent

Communications

Railways	0 km	5719 km
Highways	1296 km (paved: 416 km; unpaved: 880 km)	71,118 km (all paved)
Waterways	0 km	65 km (Rhine; open to the North Sea)
Pipelines	0 km	Crude oil 314 km; natural gas 1506 km
Airports	2 (1 for international traffic)	67 (8 for international traffic)
Telephones	Approx. 5000 units	Over 6,000,000 units

Environment

Climate	From tropical in the southern plains to extreme mountain climate in the Himalayas	Mediterranean south of the Alps; oceanic to continental north of the Alps; altitudinal dependency of temperature and precipitation
Natural resources	Timber, hydropower, gypsum, calcium carbide	Hydro-power potential, timber, salt
Land use	Arable land: 2 per cent; meadows and pastures: 5 per cent; forest and woodland: 70 per cent; other: 23 per cent	Arable land: 10 per cent; meadows and pastures: 40 per cent; forest and woodland: 26 per cent; other: 24 per cent
Environmental issues	Soil erosion; limited access to drinking water	Air pollution from vehicle emissions and open-air burning; acid rain; water pollution from increased use of agricultural fertilizers; loss of biodiversity
International agreements	Biodiversity, Climate Change, Nuclear Test Ban; signed, but not ratified – Law of the Sea	Air Pollution, Air Pollution–Nitrogen Oxides, Air Pollution – Sulphur 85, Air Pollution – Volatile Organic Compounds, Antarctic Treaty, Biodiversity, Climate Change, Endangered Species, Environmental Modification, Hazardous Wastes, Marine Dumping, Marine Life, Conservation, Nuclear Test Ban, Ozone Layer Protection, Ship Pollution, Tropical Timber 83, Wetlands, Whaling, Air Pollution – Sulphur 94, Antarctic–Environmental Protocol, Desertification, Law of the Sea, Tropical Timber 94

TABLE 1.2 Socio-economic and environmental statistics for selected countries with mountains and uplands

Country	Population (millions, 1995)	Population density (per km²)	% Urban	GDP per capita (US$, 1992)	% GDP primary sector	% GDP second sector	% GDP tertiary sector	Water resources per capita (m³)	Energy consump. per capita (MJ)
Africa									
Cameroon	6.393	28	44.9	1,029	28.6	24.9	46.5	20,252	2.72
Ethiopia	55.053	50	13.4	370	60.45	10.3	29.3	1,998	0.82
Kenya	28.261	49	27.7	914	28.9	17.6	53.5	1,069	3.18
Madagascar	14.763	25	27.1	608	33.9	13.8	52.3	22,827	1.02
Morocco	27.028	61	48.4	2,173	14.3	32.4	53.3	1,110	10.99
Rwanda	7.952	302	6.1	762	40.5	21.5	38.0	792	0.88
South Africa	41.465	34	50.8	3,068	4.6	39.5	56.0	1,206	86.29
Asia									
Afghanistan	20.141	31	20.0	800[a]	—	—	—	2,482	1.09
Bhutan	1.638	35	6.4	730	40.6	29.5	29.9	57,998	1.22
China	1221.462	127	30.3	1,493	19.5	47.6	32.9	2,292	24.29
Hong Kong	5.865	5,612	95.0	16,471	2.0	20.0	78.0	—	65.81
India	935.744	285	26.8	1,282	31.4	27.3	41.3	2,228	9.97
Indonesia	197.588	104	35.4	2,102	18.8	39.4	41.8	12,804	13.45
Iran	67.283	41	59.0	3,685	20.8	36.4	42.9	1,746	48.51
Japan	125.095	331	77.6	15,105	2	42	56	4,373	139.93
Nepal	21.918	156	13.7	1,130	43.1	21.3	35.6	7,756	0.86
Tajikistan	6.101	43	32.2	2,180	33	31	34	16,604	42.28
Turkey	61.945	79	68.8	3,807	15.1	30.4	54.5	3,117	31.94
Europe									
Albania	3.441	120	37.34	600[b]	—	—	—	6,190	12.49
Austria	7.968	95	55.52	17,690	3.0	36.0	61.0	11,333	121.23
Czech Republic	10.296	131	65.42	6,280	6.17	39.95	53.87	5,653	161.13
France	57.981	105	72.78	18,430	3.0	29.0	68.0	3,415	157.86
Georgia	5.457	78	58.45	1,640[c]	58.03	21.79	20.18	11,942	29.13
Germany	81.591	229	86.54	19,770	17.0	27.0	56.0	5,612	94.63
Greece	10.451	79	65.22	14,709	2.0	39.0	60.0	2,096	168.2
Italy	57.187	190	66.62	17,040	3.0	33.0	64.0	2,920	118.01
Norway	4.337	13	61.49	17,170	3.0	36.0	61.0	90,385	208.43
Portugal	9.823	106	35.58	9,450	9.0	37.0	54.0	7,085	61.38

Romania	22.835	96	55.39	6,900	20.51	39.88	39.61	9,109	77.16
Russian Federation	147.000	9	76.01	5,260[d]	7.9[d]	37.3[d]	54.8[d]	30,599	204.36
Spain	39.621	78	76.45	12,670	5.0	35.0	60.0	2,809	84.77
Switzerland	7.202	174	60.8	21,780	4.0	36.0	61.0	6,943	136.76
United Kingdom	58.258	239	89.46	16,340	2.0	36.0	62.0	1,219	17.83
Latin America									
Argentina	34.587	13	88.07	5,120	5.99	30.68	63.33	28,739	58.37
Bolivia	7.414	7	60.76	2,170	—	—	—	40,464	11.59
Brazil	161.790	19	78.24	5,240	10.0	39.0	51.0	42,957	23.48
Chile	14.262	19	83.9	7,060	—	—	—	32,814	37.79
Colombia	35.101	31	72.72	5,460	16.0	36.0	48.0	30,483	23.61
Costa Rica	3.424	67	49.7	3,569	15.29	25.78	58.94	27,745	18.39
Ecuador	11.460	40	58.44	2,830	12.11	37.62	50.27	27,400	21.37
Mexico	93.674	48	75.29	6,253	8.45	28.38	63.17	3,815	52.74
Peru	23.780	19	72.22	2,092	11.00	43.18	45.82	1,682	13.2
Venezuela	21.844	24	92.84	7,082	5.04	41.91	53.05	60,291	95.35
North America									
Canada	29.463	3	76.68	19,320	3.0	34.0	63.0	98,462	312.18
United States	263.250	28	76.23	22,130	2.0	29.0	69.0	9,413	310.54
Oceania									
Australia	18.088	2	84.68	16,680	3.0	31.0	66.0	18,963	216.55
New Zealand	3.575	13	86.06	13,970	8.0	27.0	65.0	91,469	158.04

[a] *L'Etat du monde* 1994, GDP for 1993

[b] Ibid., GDP for 1990

[c] Ibid., GDP for 1991

[d] *L'Etat du monde* CD ROM, GDP for 1994

TABLE 1.2 Continued

Country	Area (1,000 km²)	% Arable land	% Pasture	% Forest	% Other land	% Protected area	Endemic/known mammal species	Endemic/known bird species	Endemic/known plant species
Africa									
Cameroon	475.4	15.1	4.3	77.1	3.4	4.3	13/297	8/874	156/8,000
Ethiopia	1222.2	12.7	40.6	22.7	24.0	2.1	31/255	28/813	1,000/6,500
Kenya	580.4	7.9	37.4	29.5	25.2	6.0	21/359	6/1,068	265/6,000
Madagascar	587.0	5.3	41.3	39.9	13.5	1.9	77/105	103/253	6,500/9,000
Morocco	446.6	22.2	46.8	20.1	10.8	0.8	4/105	0/416	625/3,600
Rwanda	26.3	47.4	18.2	22.3	12.0	12.4	0/151	0/666	26/2,288
South Africa	1221.0	10.8	66.6	6.7	15.8	6.1	27/247	7/790	—/23,000
Asia									
Afghanistan	652.1	12.3	46	2.9	38.7	0.3	1/123	0/460	800/3,500
Bhutan	47.0	2.8	5.8	65.9	25.3	19.3	0/99	0/543	75/5,446
China	9597.0	10.2	42.8	13.9	32.8	3.2	77/394	67/1,244	18,000/30,000
Hong Kong	1.1	7	1	22.2	69.6	—	—/—	—/—	—/—
India	3288.0	57	3.8	23	16	4.0	44/316	55/1,219	5,000/15,000
Indonesia	1905.0	17.1	6.5	61.7	14.6	10.2	198/436	393/1,531	17,500/27,500
Iran	1648.0	11	26.8	6.9	55	4.8	5/140	1/502	—/—
Japan	377.8	11.8	1.7	66.6	19.7	12.3	38/132	21/583	2,000/4,700
Nepal	140.8	17.2	14.6	42	26.1	7.9	1/167	2/824	315/6,500
Tajikistan	143.0	6	25.2	3.8	64.9	0.6	2/—	0/—	—/—
Turkey	779.5	35.7	16	26.2	21.8	0.3	1/116	0/418	2,675/8,472
Europe									
Albania	28.8	25.6	15.4	38.2	20.8	1.5	68/0	306/0	2,965/24
Austria	83.9	18.1	23.6	39.1	19.1	25.3	83/0	414/0	2,950/35
Czech Republic	78.7	42.6	11.2	34.0	12.1	—	—/0	—/0	—/—
France	551.5	35.3	19.5	27.1	18.0	9.6	93/0	506/9	4,500/133
Georgia	700.0	14.3	26.8	38.7	20.0	2.7	—/2	—/0	—/—
Germany	356.9	34.6	15.0	30.6	19.9	24.6	76/0	503/0	2,600/6
Greece	132.0	27.1	40.7	20.3	11.8	0.8	95/2	398/0	4,900/742
Italy	301.3	40.3	14.6	23.0	22.0	6.7	90/3	490/0	5,463/712
Norway	323.9	2.9	0.4	27.1	69.5	5.0	54/0	453/0	1,650/1
Portugal	92.4	34.3	9.1	35.8	20.6	6.1	63/1	441/2	2,500/150
Romania	237.5	43.1	21.0	29.0	6.8	4.6	84/0	368/0	3,175/41

Russian Federation	1708.0	7.8	4.4	45.8	42.0	1.2	—/—	—/—	—/—
Spain	504.8	39.3	20.6	32.3	7.7	6.9	82/4	506/5	—/—
Switzerland	41.3	11.8	28.1	31.6	28.4	18.2	75/0	400/0	1650/1
United Kingdom	244.9	25.3	45.7	10.1	18.8	18.9	50/0	590/1	1550/16
Latin America									
Argentina	2767.0	9.9	51.8	18.5	19.5	3.4	320/47	976/19	9,000/1,100
Bolivia	1099.0	2.1	24.4	53.4	19.8	8.4	316/20	1,274/16	16,500/4,000
Brazil	8512.0	5.7	21.8	57.7	14.6	3.3	394/96	1,635/177	55,000/—
Chile	757.0	5.6	18.1	22.0	54.1	18.1	91/16	448/15	5,125/2,698
Colombia	1139.0	5.2	39.0	48.1	7.5	8.2	359/28	1,695/62	50,000/1,500
Costa Rica	51.1	10.3	45.8	30.7	13.0	12.1	205/6	850/7	11,000/950
Ecuador	283.6	10.9	7.5	56.3	25.1	39.3	302/23	1,559/37	18,250/4,000
Mexico	1958.0	12.9	39.0	25.5	22.4	5.1	450/140	1,026/89	25,000/12,500
Peru	1285.0	2.6	21.1	66.2	9.9	3.2	344/45	1,678/109	17,121/5,356
Venezuela	912.1	4.4	20.1	34.0	41.4	30.2	305/16	1,296/42	20,000/8,000
North America									
Canada	9976.0	4.9	3.0	53.5	38.4	5.0	193/7	578/3	2,920/147
United States	9373.0	19.6	24.9	29.8	25.5	10.5	428/101	768/70	16,302/4,036
Oceania									
Australia	7713.0	6.0	54.1	18.9	20.8	10.6	252/198	751/353	15,000/14,074
New Zealand	271.0	14.1	50.3	27.5	7.9	10.7	10/4	287/76	2,160/1,942

[a] *L'Etat du monde* 1994, GDP for 1993

[b] Ibid., GDP for 1990

[c] Ibid., GDP for 1991

[d] *L'Etat du monde* CD ROM, GDP for 1994

various spatio-temporal scales. Erosion is a constant feature of mountain environments, from the long-term effects of glaciological, hydrological and chemical weathering (which act on time scales from centuries to millennia or more), to the sudden manifestation of major natural catastrophes such as rock falls and mud slides (which may occur over a time-span of but a few minutes). The extent to which a region will be sensitive to the agents of erosion depends largely upon the geology of the region considered, its climate, and associated factors such as slope and exposure to weathering elements.

Whatever the source of the disturbance, mountains are composed of a number of inherently fragile systems which have difficulty in adapting to changing environmental conditions. It may appear paradoxical that mountain plants, for example, which are capable of resisting extreme climatic conditions, may be threatened by extinction following a seemingly minor environmental stress. This can be explained by the particular characteristics of the mountain environments themselves, which all contribute to, and interact with, vegetation processes. Plants have adapted to climatic extremes, short growing seasons, lack of soil nutrients, steepness of slopes, competition between species, etc. These conditions allow plants to survive within a very narrow 'environmental bandwidth', and any disturbance to their basic living conditions can lead to severe stresses to particular types of vegetation. Perhaps the elements most sensitive to environmental change in the mountains are snow and ice; if climatic conditions become warmer, then the only possible response of snow fields and glaciers is to melt. In this respect, the mountain cryosphere is a valuable indicator of climatic change.

Humankind is adding a new dimension to the global environment in general, and mountain regions are no exception. 'Human interference' is generally perceived as being detrimental to mountain environments; while this may often be the case, there are examples where human activities have augmented the safety or the aesthetics of mountains. Terraced agriculture as practised in many parts of the world (Nepal, Indonesia, the Philippines) is considered by many to enhance the beauty of a region; in addition, this form of land use acts to stabilize mountain slopes.

In other regions, however, direct or indirect environmental mismanagement has led to irreversible damage: notable examples include the denudation of mountain slopes following forest fires, as in parts of the Mediterranean basin or in California, the dieback of mountain forests as a result of industrial pollution (mountain regions of Poland, the Czech Republic and Germany), and open-pit mining such as that practised in Chile or the United States. Environmental considerations are often opposed to economic interests; an example typical of such potential sources of conflict is the construction of hydro-power facilities, which sometimes destroy unique ecosystems or lead to the displacement of indigenous people; a notable example is the mega-project of the 'Three Rivers' in China, in which a 600-km artificial lake will lead to the displacement of an estimated 2 million people. In partially taming the mountain world, our species has often generated as many problems as it has resolved. The example provided in Box 1.1 related to transportation across the Alps is an illustration of this dichotomy.

1.5 Global environmental change: fundamental issues

Global environmental change can be defined as a series of stress factors on the physical and biological systems of the planet. The Earth's environment is continuously subjected to various stresses through natural processes and human interference. Global change is not a new concept, but with the rapid industrialization and population growth which the twentieth century has witnessed worldwide, the natural environment has undergone unprecedented changes. In some instances, environmental degradation is inevitable because of the basic requirements of human populations, particularly where those are growing rapidly; in other cases, environmental damage is a direct result of mismanagement and overexploitation

Box 1.1 Communication routes across the Alps

Since the Middle Ages, traffic crossing the Alps has been important for the mountain regions themselves as well as for the economic centres for which it provides trade links. Originally a series of arduous mule paths (some now renovated to attract tourists), today the major routes consist of a combination of pass roads and tunnels for rail or road, the mountains no longer representing the same physical barrier (Burri, 1995; Swiss Federal Office of Transportation, 1991).

The ability to traverse the Alps more easily has multiple advantages in terms of well-being and economic development: increased trade between the regions on either side, maintaining long-distance social contacts, enjoying year-round leisure activities in the mountains, etc. Economically poor regions in the Alps have benefited considerably from transit and tourist traffic. In the decades following the Second World War, however, the rapid increase in passenger and goods transport (tripling in the last 25 years) has led to evident negative impacts, in particular noise, air pollution, greenhouse gas emissions, landscape degradation, congestion and accidents (Swiss Federal Statistical Office, 1997). Motorized vehicles are the major source of these problems, which are exacerbated by the Alpine topography's steep slopes and narrow valleys.

The distribution of different modes of transit traffic demonstrates the effects that transportation policy can produce. In Switzerland, 82 per cent of freight crossing the Alps was carried via rail in 1995, as opposed to 22 per cent in France and 30 per cent in Austria. This results mainly from Switzerland's restrictive policy regarding truck transport, banning travel on Sundays and at night, and setting a load limit of 28 tons (Swiss Federal Statistical Office, 1997). Though the aim of these measures was to enhance rail's competitive position in

relation to road transport in Switzerland, the consequences are felt strongly in France and Austria. In these countries, an estimated 30 to 40 per cent of heavy vehicle traffic expected to traverse the Alps through Switzerland is rerouted through France and Austria, displacing the pollution, noise and congestion problems associated with it.

Three major policies promoting train transportation have been approved by Swiss voters: 'Rail 2000', 'New Transalpine Railway Axes' (NEAT, in the German acronym) and the 'Alps Initiative'. The latter is a constitutional amendment requiring that, by 2004, all transit freight traffic be shifted to rail within Swiss borders (Burri, 1995). Significantly, it was launched by the residents of those cantons most affected by the negative consequences of transit traffic (Uri, Valais, Schwyz and Ticino; Mayer-Tasch et al., 1990). NEAT provides for the construction of new, longer and deeper 'base' rail tunnels through the Gotthard and Lötschberg massifs (Burri, 1995). Implementing these policies has proven to be a complex matter; aside from obstacles such as the geology of the Gotthard region and the often political questions involved in scheduling, financing these projects remains a principal issue (Jochimsen and Kirchgässner, 1995).

Efforts are being made to reduce subsidies for public transportation and, at the same time, to incorporate the external costs of motor vehicle traffic into the prices paid by road users. For 1993, external health, accident and environmental costs together with the costs of accidents and noise of overall transport are estimated at 4.1 billion Swiss francs per annum. Road transport accounts for about 85 per cent of these problems. The 'Alps Initiative' includes steps to internalize these costs with a distance-based charge on heavy goods vehicles and an alpine transit charge (Swiss Federal Statistical Office, 1997).

Obviously, the trans-Alpine traffic debate is a trans-boundary one requiring supranational cooperation. Bilateral negotiations between Switzerland and the European Union in 1998 overcame numerous hurdles dealing with transport issues. Swiss negotiators were in the difficult position of having to provide for the implementation of national legislation encouraging rail use on the one hand, and considering the needs of Switzerland's European neighbours who have a 40-ton load limit, on the other hand. A gradual increase of the weight limit by Switzerland was agreed to in exchange for a gradual introduction of the distance-dependent charge for transit trucks (Swiss Federal Statistical Office, 1997). The main hindrance to closing the talks involved agreement on the transit charge introduced by Switzerland. However, closure of negotiations has not ended the debate; for example, environmental organizations find the tax too low to shift road traffic to the railways and demand additional measures, such as the raising of fuel prices. Austria, which must accept considerable road traffic on the Brenner Pass into Italy, has reacted by insisting that transit traffic not be charged significantly less to cross the Alps there.

An innovative short-term solution to the problem of transit truck traffic is offered by Switzerland's 'road–rail' service, loading trucks on rail carriers to transport them from border to border. This fulfils one of Switzerland's transportation policy goals of modal combination. For the long term, a Europe-wide system of container transport is considered a desirable goal for which Switzerland is willing to contribute to the financing for construction of transfer centres located outside the country (Swiss Federal Office of Transportation, 1991).

of natural resources. The consequences of such degradation are sometimes not recognized or are ignored because of the perceived higher benefits of economic gain.

Whether the global environment is capable of withstanding natural and anthropogenic stresses is a matter of constant debate. In some instances, the environment has – at least on local to regional scales – been able to revert to its previous levels. Examples of such resilience abound; for example, the acidity of lakes in northern Ontario, Canada, reverted to its natural levels following significant abatement of sulphur-based pollution from a major smelting plant (Gunn and Keller, 1990). In other instances, environmental damage appears to be irreversible, such as the large perimeter of contaminated lands following the 1986 nuclear accident in Chernobyl, Ukraine, or the deforestation of pristine jungles in many tropical regions. Examples of irreversible degradation have provided arguments to those who believe that environmental impacts are cumulative and difficult to reverse. A third paradigm is that there are certain beneficial effects of stresses on the environment, namely that ecosystems become resilient and can therefore withstand further and possibly greater stresses in the future. Forest fires, for example, are not solely a devastating phenomenon: they also return essential nutrients to the soil, allowing vigorous regeneration to occur. Indeed, some environments can maintain themselves in the long term only through fire.

All these paradigms are based on the assumption that the time scales associated with environmental change are long and that, in many situations, the environment may find a new equilibrium, if not its original state. Environmental upheavals have occurred in the past, along with species extinctions, and yet the planet has 'survived' and evolution has continued. However, it is possible that anthropogenic pressures are accelerating change and that many systems may not adapt to rapid rates of change, even if they could adapt to the amplitude of change over longer time periods.

Under the heading of global environmental change, one can list the following contributing factors, many of which are essentially human-induced:

• air pollution and ozone depletion

- climatic change
- land degradation
- deforestation
- desertification
- loss of biodiversity
- availability of fresh water
- hazardous wastes
- war.

There are two principal causal mechanisms which can account for human interference on the natural environment: economic growth and demography. The economic level of a country determines to a large extent its resource requirements, in particular energy, industrial commodities, agricultural products and supply of fresh water. Demography, on the other hand, is a critical factor in the sharing of the resources available to a particular country or set of countries.

High economic levels are resource-intensive, and this frequently leads to environmental degradation because of the resources required to maintain a high standard of living. As an extreme example, Table 1.1 shows that the energy use per capita in the United States is 350 times greater than in Ethiopia or Rwanda. Technology is today still energy-intensive, particularly in the transportation sector, where fuel demand for road and air transportation continues to grow rapidly. Technology as used in the industrialized countries, along with the image of 'Western lifestyles', is often replacing traditional consumption and resource-use patterns in the developing world, even though traditional methods are often better adapted to local conditions. As a consequence, environmental concerns become a very low priority as economic survival becomes the dominant objective.

High population growth, on the other hand, can also lead to environmental damage as the inhabitants of poor countries attempt to maintain or improve their current economic level through the exploitation of their resources, often without any long-term planning or management. While in most of the industrialized world, population growth is low (often less than 1 per cent), in some developing countries, demography is such that economic levels are

dwindling. Kenya, for example, has a rate of population growth of around 4.3 per cent per annum; if this is sustained, its population will double within a time-frame of 17 years. In order just to maintain present economic levels, Kenya would need to double its existing infrastructure (energy supply, housing, schools, hospitals) and food and water supply. Such pressing demands on an already depressed economy represent a challenge which even the most affluent countries would not be able to meet. The general lowering of a population's standard of living means that economic survival takes place at the expense of environmental protection, management and planning.

The current trends of globalization of economic markets and highly liberalized economies are not always compatible with environmental concerns. Present-day economic policies are often based on the short term (from a few days to a few months), whereas environmental management is in essence based on the long term (several years to several decades). Environmental protection is sometimes perceived to be contrary to free trade; indeed, attempts to slow down and possibly reverse tropical deforestation are considered in some circles to be contravening the GATT and WTO accords on free trade. (GATT was the General Agreement on Tariffs and Trade; its successor is the World Trade Organization or WTO.) In the context of economic recession as experienced in many industrialized countries in the 1990s, environmental protection is sometimes seen as a menace to job security (i.e. environmental concerns = obstacle to a particular industry or business = higher fiscal pressures to contribute towards environmental protection = loss of jobs to compensate for increased financial expenditure). However, environmental assessments leading to protection and reclaiming have also led, in some instances, to new economic activities and employment opportunities.

Because many actors in the economic and industrial arena tend to believe that the natural environment has no intrinsic value, there is often little incentive towards environmental protection or management, since the environment is perceived as an unlimited resource to be utilized in order to sustain economic growth. In the United States, there is increasing pressure from industrial lobbies for access to resources in some of the national monuments and national parks. There is the fear by environmentalists that allowing even limited rights for mining and oil prospection in protected areas such as Death Valley National Park (California), or in the more recent parks of Alaska, could set a precedent whereby the protection of these parks would become meaningless at some point in the future. Exploitation of raw materials by major manufacturing firms is a direct threat to indigenous people in different parts of the world; it is estimated that about 300 million people live in regions which account for about 60 per cent of the world's natural resources.

Globalization is leading to a real or perceived trend of weakening of policy-making at the national level. Because major economic and financial decisions are taken outside a purely national framework, the ability of politicians to respond to such decisions is reduced. Politics, which up till recently was the driving force behind the nation-state, is today increasingly involved in economic management. So powerful is the wave of globalization that many policy-makers are more concerned with economic affairs than with environmental issues. They are also sensitive to shifts in the priorities of the general public, which is today more preoccupied by economic conditions than by the environment. In the industrialized countries, unemployment, crime and health insurance are all considered more pressing issues than those pertaining to environmental matters. In the developing world, food security and access to basic commodities and health are perceived as being far more urgent issues than a healthy environment. Conflicts between environmental issues and economic development have been described and understood for a long time. Awareness of these issues has progressively resulted in the concept of sustainable development.

Whatever the perception, it is highly likely that global environmental change will lead to lasting degradation and damage. This will in turn reduce the capacity of human societies to maintain their lifestyles at current levels, in particular because the driving forces of global economy may no longer be able to use in a sustained manner the resources which the environment provides. Regional disparities between rich and poor, which have always existed, will be exacerbated in a degraded environment. Issues related to global change highlight the general problem that society has difficulty in using the limited resources of the planet in a rational manner, and of sharing equitably the essential commodities which the environment provides.

2

Characterization of mountain environments

2.0 Chapter summary

This chapter provides an overview of different aspects of physical and human systems in mountain regions as they are perceived today. There is a strong bias towards climate and atmospheric processes, because climate is one of the dominant environmental controls on natural systems in mountains. Climate determines to a large degree the amount and timing of runoff in the main hydrological basins of the world; it plays a key role in snow and ice formation; and it exerts a major influence on the distribution of plant species and ecosystems, both altitudinally and latitudinally. The major interactions between mountains and climate, from the global down to the local scales, are reviewed; there is a subsequent overview of specific issues related to mountain hydrology, cryosphere, soils and vegetation, as well as the interlinks between all these systems.

Problems related to the human presence in mountains and uplands, in both the developed and the developing world, are discussed towards the end of the chapter. Issues of health, tourism and resource use are also addressed. Distinctions are made for the level of economic activity taking place in different mountain regions of the world, as well as the environmental impacts which these imply.

2.1 Climate

Climate is the principal factor governing the natural environment on short time scales (as against plate tectonics, which is responsible for the formation of mountains). Climate characterizes the location and intensity of biological, physical and chemical processes. It is for this reason that a large part of the present chapter will be devoted to the climatic component of the environment. Mountain climates consist of four major factors (Barry, 1994), which will be addressed in the following subsections, namely:

- continentality
- latitude
- altitude
- topography.

2.1.1 Continentality

Continentality refers to the proximity of a particular region to an ocean. The diurnal and annual ranges of temperatures in a maritime climate are markedly less than in regions far removed from the oceans; this is essentially due to the large thermal capacity of the sea, which warms and cools far less rapidly than land. Because the ocean represents a large source of moisture, there is also more precipitation in a maritime climate than in a continental one, provided the dominant wind direction is onshore. Examples of maritime mountain climates include the Cascade Ranges in Oregon and Washington States of the USA, the Alaskan coastal mountains, the New Zealand Alps, the Norwegian Alps and the southern Chilean Andes. Mountains under the dominant influence of continental-type climates include the Tibetan Plateau, the mountains of Central Asia (Pamir, Tien Shan, the Urals) and the Rocky Mountains in Colorado, Wyoming and Montana. However, many other mountain regions often define and separate climatic regions; for example, the European Alps act as a boundary between Mediterranean-type climates and Atlantic and continental climates to the north.

Table 2.1 illustrates typical examples of continental and maritime climates for Sverdlovsk (foothills of the Ural Mountains, Russian Federation) and Bergen (on the coastal ranges of Hordaland Province, Norway). The influence of the ocean on both temperature range and precipitation is quite obvious. There is no month where mean monthly temperatures are below freezing point in Bergen, despite the fact that it is located above 60° N. Precipitation is distributed fairly evenly throughout the year, with an annual total of almost 2000 mm. In Sverdlovsk, on the other hand, the annual temperature range is close to three times that of Bergen, for just over one quarter of the annual precipitation. Precipitation tends to fall more during the summer months, in the form of convective downpours. Winter months are extremely cold and very dry, with less than 30 mm of precipitation per month from October to May, mostly in the form of snow.

While mountains in continental regions experience more sunshine, less precipitation and a larger range of temperatures than maritime mountains, they are not necessarily harsher environments. For certain ecosystems, the larger amounts of sunshine compensate for lower mean temperatures in continental regions (Price, 1981). Increased cloudiness and precipitation (both rain and snow) in coastal mountain ranges such as the Cascades or the New Zealand Alps limit the growth of certain species despite the milder overall temperatures. The timber line in continental

TABLE 2.1 Statistics for two sites characteristic of continental and maritime climatic regimes respectively

	Sverdlovsk		Bergen	
	Temperature (°C)	Precipitation (mm) (Scudden)	Temperature (°C)	Precipitation (mm)
January	−14.6	15	1.5	179
February	−13.4	17	1.3	139
March	−7.5	17	3.1	109
April	3.3	22	5.8	140
May	10.3	40	10.2	83
June	16.4	59	12.6	126
July	17.8	80	15.0	141
August	15.8	82	14.7	167
September	9.4	49	12.0	228
October	1.9	29	8.3	236
November	−7.1	25	5.5	207
December	−13.0	27	3.3	203
Mean	1.6		7.8	
Range	32.4		13.7	
Total		462		1,958

Sverdlovsk (56.8° N, 60.6° E) is 237 m above sea level; Bergen (60.4° N, 5.3° E) is 43 m above sea level regions is often located at higher elevations than in maritime zones, which confirms the importance of these compensating factors for regional ecological systems (Wade and McVean, 1969; Wardle, 1973).

2.1.2 Latitude

Latitude in mountains, as in other regions, determines to a large extent the amplitude of the annual cycle of temperature and, to a lesser extent (because of local site characteristics and continentality), the amount of precipitation which a region experiences. Mountains tend to amplify some of the characteristics of tropical, mid-latitude and boreal climates for reasons related to topography (*see* Section 2.1.4).

Figure 2.1 illustrates the climatology of three mountain regions characteristic of the tropical, temperate and boreal zones, respectively Mt Kenya (Africa), Säntis (Switzerland) and Mt Denali (Alaska). It is seen that as one proceeds equatorwards, the amplitude of the annual temperature cycle is reduced (from a maximum of 37°C in Alaska, in this example, to about 1°C in Kenya). Precipitation, on the other hand, reflects not only the latitudinal characteristics but also regional particularities. In Kenya, precipitation is influenced in April and May by a monsoon-type circulation, bringing the 'long-rains' period, while the 'short rains' are a secondary occurrence of precipitation in November and December. In Switzerland, precipitation falls throughout the year, with a relative maximum in the summer

months. In the vicinity of Mt Denali, summer precipitation is clearly the most intense, in the form of summer showers, while winter snowfall is rather sparse because of the very cold conditions prevailing from November to April.

2.1.3 Altitude

Altitude is perhaps the most distinguishing and fundamental characteristic of mountain climates. Atmospheric density and pressure decrease with height in the troposphere (which is the region within which even the highest mountains are confined), as does temperature, because the lower the density of the air, the lower its ability to retain heat. Figure 2.2a illustrates the average altitudinal profile of pressure; it is seen that the height-pressure relationship is not linear. Despite the fact that water vapour represents less than 5 per cent of the composition of the atmosphere, it is one of the few climatically relevant gases, along with a number of so-called 'greenhouse gases' which are present in trace quantities. The radiative properties of water vapour exert a major control on the thermal structure of the atmosphere. However, vapour is present essentially in the layers close to the surface, with about half the vapour content of the air confined to the lowermost 2000 m of the atmosphere; above this level, water vapour diminishes rapidly, as seen in Figure 2.2b.

At high elevations, therefore, thermal conditions are often extreme; the only source of energy is direct solar radiation that is absorbed

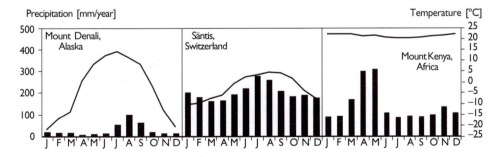

FIGURE 2.1 Statistics of annual temperature (solid line) and precipitation (histograms) for characteristic boreal, temperate and tropical mountain ranges

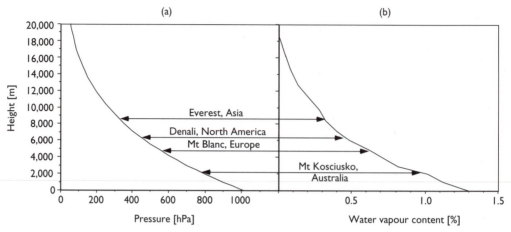

Figure 2.2 (a) Vertical profile of atmospheric pressure; (b) vertical profile of atmospheric water vapour content. Altitudes of several major mountains are illustrated to provide an idea of their vertical dimension and the implications this has for environmental systems

by the surface. Mountain soils can heat up rapidly during the day, and exchange some of this heat with the air in contact with the surface, but because of the lack of air and water vapour molecules, only a very thin envelope of air is likely to be warmed. This leads to steep vertical temperature gradients in the first few metres above the ground. Mountains in this sense serve as elevated heat sources, whereby diurnal temperatures are higher than at similar altitudes in the free atmosphere (Flohn, 1968). Diurnal and annual range of temperature tends to decrease with altitude because of the lower heat capacity of the atmosphere at height.

The altitudinal controls on mountain climates also exert a significant influence on the distribution of ecosystems. Indeed, there is such a close link between mountain vegetation and climate that vegetation-belt typology has been extensively used to define climatic zones and their altitudinal and latitudinal transitions (e.g. Klötzli, 1984, 1991, 1994; Ozenda, 1985; Quezel and Barbero, 1990; Rameau *et al.*, 1993; *see* Section 2.5).

2.1.4 Topography

Mountain systems generate their own climates (Ekhart, 1948), as a function of the size of the land mass at a particular elevation. Topographical features also play a key role in determining local climates, in particular the slope, aspect, and exposure of the surface to climatic elements. These factors tend to govern the redistribution of solar energy as it is intercepted at the surface, as well as precipitation which is highly sensitive to local site characteristics.

2.1.4.1 Orographic forcing

In many low- and mid-latitude regions, precipitation is observed to increase with height; even modest topographic elements can exert an often disproportionate influence on precipitation amount. Precipitation mechanisms are linked to atmospheric dynamics and thermodynamics. When a mass of moist air is uplifted above the condensation level, water vapour exceeds a critical threshold beyond which air at a given temperature can no longer hold water in its vapour form. The air is said to reach its dew point, or condensation level; it thus becomes saturated in the presence of condensation nuclei, and the excess vapour is converted to fine liquid water particles that become visible in the form of mist, fog or clouds. If uplift of air continues, there can at some stage be sufficient liquid cloud droplets (or ice crystals if the ambient temperature is below freezing) for coalescence processes to occur, i.e. the aggregation of numerous droplets or crystals into much larger raindrops or snowflakes. Beyond a given size, gravity will overcome buoyancy forces, thus enabling these drops or snowflakes to fall out towards the surface.

In the free atmosphere, uplift of air occurs as a result of thermal instability or of frontal activity linked to cyclonic storm systems. In the former case, surface temperature heterogeneities can force warmer, lighter air to rise, entraining with it moisture which may ultimately condense and form precipitation; in the case of frontal activity, uplift occurs when two air masses meet and dynamically force warmer, less dense air above colder, denser air. In mountain regions, on the other hand, uplift will occur whenever an airflow trajectory intersects an orographic barrier. The dynamics of the flow will force the air either around the barrier or above it; in the latter case, the orographically forced motion may entrain sufficient moisture upwards for condensation and precipitation processes to come into effect. Forced ascent is most effective when the flow of air is perpendicular to a mountain barrier; also, the steeper the slope and the greater its exposure, the faster will be the uplift.

The cloud types associated with uplift, generically known as convective clouds (cumulus or cumulonimbus), occur both in the presence and in the absence of mountains, but because of the physical forcing by a mountain barrier, convective processes are more effective in the presence of mountains. Precipitation is by association also enhanced by orography; some of the highest precipitation systematically recorded occurs on the Big Island of Hawaii (Waialeale), which receives annually close to 12,000 mm of rain. Cherrapunji, India, located in the foothills of the Himalayas and relatively close to the Bay of Bengal, receives over 11,000 mm, mostly during the intense monsoon season.

2.1.4.2 Rain shadows

Precipitation in a mountain region will generally fall out in the windward-facing slopes of the mountains because of the dynamics associated with uplift. Because most of the moisture is extracted from the clouds on the windward slopes, the air which enters the lee side of the mountains is mainly dry. Clouds may form and appear to adhere to the mountain crests, but little precipitation is likely to fall on the leeward sides of the mountains. Death Valley in California, one of the most arid desert regions in the world, is located on the leeward side of the highest part of the Californian Sierra Nevada mountains, which rise to 4417 m at Mt Whitney. Indeed, most of the desert regions of the western United States lie to the east of the coastal ranges which stretch from southern California to northern Washington. Precipitation gradients are steep from one side of the ranges to the other; for example, San Francisco, California, receives an average of 475 mm of precipitation annually (double or triple the amount in the upper reaches of the Sierra Nevada), while Las Vegas, Nevada, receives only 99 mm. On more local scales, the windward and leeward influences on precipitation are particularly marked on islands which have a significant orography; the main city of Big Island, Hawaii (Hilo), experiences close to 3500 mm annually, while the resort areas of Kailua-Kona, situated in the lee of the 4000-m Mauna Laua and Mauna Kea volcanoes, receive a sparse 65 mm annually.

2.1.4.3 Lee cyclogenesis

Lee cyclogenesis is the process by which active frontal systems are formed downwind of mountain barriers. Because mountains can penetrate the atmosphere to great heights, they encounter strong airflows; mountain crests and tops can therefore experience extremely high winds, particularly in mid-latitudes, where the dynamics of the general circulation of the atmosphere tends to be most intense. Through their physical presence, mountains modify atmospheric flows, and their influence is felt on scales which are many times their height in both the horizontal and vertical directions. High mountain chains such as the Himalayas influence the location and intensity of the jet streams, which in turn determine climatic

Box 2.1   Föhn-type winds

The thermodynamics associated with uplift, condensation and precipitation can, under certain circumstances, lead to wind systems which are found only in mountain regions: the föhn winds. These are found in many mountain regions of the world, but are best documented in the Alps (from where the name 'föhn' originates) and the Rocky Mountains (where a similar wind is referred to as the chinook).

Dry air moving upwards will expand with height as the density of the atmospheric environment diminishes. The so-called dry adiabatic lapse rate is a measure of the rate of change with temperature as dry air expands or contracts; as it moves up or down, air will undergo changes of 9.8°C/km. When condensation processes occur, however, latent heat energy is released, so that when uplift continues, cooling of the upward-moving air will occur at a different rate according to the amount of moisture present. This is known as the saturated adiabatic lapse rate and can range from 5°C/km to 7°C/km.

Figure 2.3 illustrates the thermodynamic cycle associated with the föhn or chinook winds. Air on the windward side of the mountains (A) undergoes uplift and cools at the dry adiabatic lapse rate until it reaches the condensation point (B); here, clouds will form and the upward motion will continue, but at the saturated adiabatic lapse rate. When precipitation occurs, moisture is removed from the clouds; with less and less moisture present, the thermodynamic trajectory of the upward-moving air tends asymptotically towards the dry adiabat, thereby explaining the non-linear aspect of the saturated ascent curve. When the air reaches the summit of the mountain range (C), it will begin to descend in the lee of the mountains. The air is now very dry, since it has lost most of or all its moisture through precipitation on the windward slopes. The downward-moving air will warm through compression at the dry adiabatic lapse rate, so that when it reaches the base of the leeward slopes (D), it will be considerably warmer than the air at the same level on the windward side of the range.

In addition to the temperature cycle associated with the föhn or chinook, there is also a vigorous dynamic component to it. The downward-flowing air will penetrate valleys to the lee of the mountains at high speeds, often in excess of 40–50 m/s. These winds can generate structural damage and, because of the high-frequency

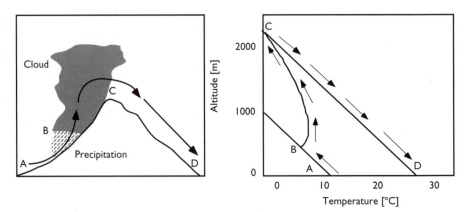

FIGURE 2.3 Schematic illustration of the thermodynamic mechanisms associated with a föhn wind

micro-pressure fluctuations related to the high turbulence, cause physiological stress to humans.

In the Alps, föhn winds are associated with southerly or northerly flows, and occur throughout the year, with a peak frequency in late fall and early spring. A southerly föhn episode can release large quantities of snow on the southern slope of the Alps, while simultaneously melting snow on the northern slopes and valleys because of the sudden rise in low-level temperatures.

conditions at locations far removed from the mountains themselves.

Disturbance of atmospheric flows by mountains generates wave-like patterns similar to the wake of a ship in the ocean. This can result in the formation of a high-pressure ridge upwind of the mountains (essentially through compression of air against the mountain barrier), while downwind, a closed centre of low pressure may exist. So long as the dominant airflow is maintained across the mountains, the position of the orographic low pressure will remain relatively stationary; its cloud and precipitation activity will not be very important because of the downward component of the air in the lee of the mountains. However, the low can be activated by the passage of a frontal system, which contributes to a deepening of the low-pressure system and its transformation to a fully-fledged frontal wave depression.

Lee cyclogenesis occurs in many mountain regions and is responsible for potentially severe weather; lows to the east of the Rockies (e.g. in Colorado) sometime force warm, moist air from the Gulf of Mexico to come into contact with cold, Arctic air originating in Canada. Such situations can lead to strong storms at the air-mass interface, sometimes accompanied by devastating tornadoes east of the Rockies. In the Alps, one of the most spectacular manifestations of lee cyclogenesis occurs in the Gulf of Genoa, as a result of westerly flow crossing the southern part of the French Alps (which have a north–south orientation and are consequently at right angles to the flow). The Gulf of Genoa low is often responsible for heavy precipitation in the north Italian plains and the southern slopes of the Alps, and frequent outbreaks of strong föhn episodes.

2.1.4.4 Gravity wave drag

The transfer of momentum between the atmosphere and the Earth occurs at all spatial scales. At the small end of the scale spectrum, dissipation of energy occurs through friction with the surface. At larger scales, additional processes lead to momentum dissipation, through the conversion of kinetic into thermal energy, and these are essentially linked to the presence of orography. Mountains which are present over large horizontal distances and with significant heights may generate internal gravity waves, which results in gravity wave drag, as sketched in Figure 2.4.

Gravity waves generated by the presence of underlying orography are capable of breaking in a similar manner to ocean waves at the seashore, and in doing so transfer substantial quantities of momentum from the large-scale to the small-scale flows. They also have an influence on the formation of clouds and precipitation close to mountains; such features contribute to heat and moisture transfer in the vertical. The major mountain ranges of the world produce large-amplitude waves that extend around the globe, and help propagate energy from one part of the world to another, particularly in the form of mid-latitude frontal systems.

2.1.4.5 Blocking

The presence of major mountain systems is one of the trigger mechanisms for episodes of persistent high-pressure cells, known as blocking highs, which can occur for periods of several days to several weeks in different parts of the world. Blocking highs over Europe are linked to the presence of the Alps, which affect the general circulation patterns over the North

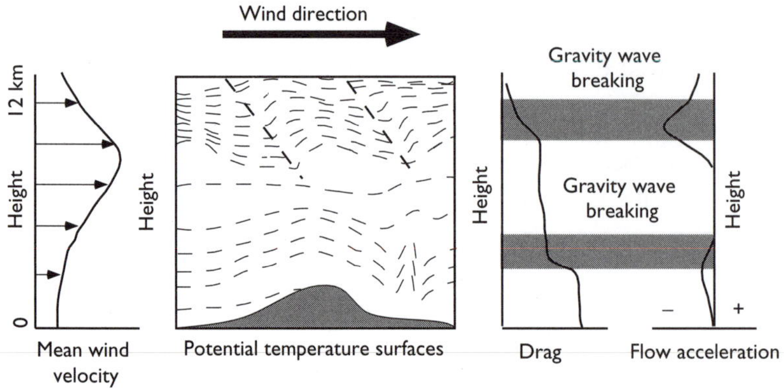

FIGURE 2.4 Illustration of the dynamic effects of a mountain barrier on the large-scale flow

Atlantic and lead to high-pressure ridges over Western Europe. According to the strength of the westerlies, the high-pressure ridge can amplify and result in a persistent anticyclonic field which may stretch across much of Western and Central Europe. Storm tracks are diverted towards Scandinavia and cannot penetrate into the western and central parts of the continent.

Blocking highs frequently occur in winter because of the cold air centred on the mountains. High-pressure cells are accompanied by anomalously warm temperatures and little or no precipitation during crucial periods of the winter season. If snow does not fall abundantly prior to the setting in of a persistent high-pressure system, then the ski season in regions such as the Alps may be affected by a thin or non-existent snow-pack. The Alps experienced an unusual run of mild winters with less than average snow conditions in the late 1980s and early 1990s that led to considerable economic hardship for resorts which depend almost exclusively on winter tourism to drive the local economy (Beniston *et al.*, 1994).

2.1.4.6 Mountain and valley breezes

Solar radiation is intercepted differently according to the orientation of slopes. In the mid- and high latitudes of the northern hemisphere, slopes oriented towards the south receive more energy per unit area and therefore experience a larger thermal amplitude than slopes with different orientations. Differential absorption and distribution of energy at the surface leads to different atmospheric responses, because air in contact with a warm surface tends to rise, as opposed to air in the vicinity of a colder surface. Air at the same elevation in the centre of a valley also warms less rapidly than the air in direct contact with the ground, so that the density differences from one side of the valley to another are capable of driving local mountain and valley breezes. Because less dense air moves up the valley slopes, it needs to be replaced by air from lower down in the valley; during the day, therefore, flow close to the valley floor is directed up-valley. The reverse situation occurs at night, when energy at the surface is lost through infrared radiation and cools the air in contact with the surface; continuity principles imply that the air moving downward toward the valley floor needs to be evacuated, so that nocturnal flows are directed down-valley.

This cyclic pattern of up- and down-valley air is present in all mountain regions, and each valley can be considered to have its own unique microclimate. Mountain and valley breezes are important in industrialized or urbanized valleys because of their effect on air quality through the recycling and dispersal of pollutants of local origin.

2.1.4.7 Temperature inversions

At night, or during periods of persistently unperturbed weather, as can occur during a stable period of high pressure, temperature inversions form as a result of high energy loss by infrared terrestrial radiation. Under such circumstances, cold air accumulates at the bottom of a valley floor and tends to stagnate because of the constraining effects of the valley boundaries. The depth of the inversion depends on local weather and topographic characteristics, but in general does not exceed 1000 m. Temperatures above the inversion are milder, because the colder, denser air flows to the valley floor and remains there. It is only at much higher elevations that the general rule of decreasing temperatures with height is verified once again. According to local site characteristics, i.e. valley floor, mountain slope or mountain top, temperatures at equivalent elevations can be vastly different from one another. This has significant implications for the distribution of climate-sensitive systems, such as vegetation, snow and ice.

2.2 Hydrological systems

The sources of all the world's major rivers are located in high mountain areas; they provide the principal water catchment areas, or watersheds, for much larger geographical regions. Rainfall is usually greater at middle altitudes than in adjoining lowlands or at the highest elevations (see Section 2.1.4.1), and water from mountains provides the source of most rivers and streams on which lowland peoples depend for water supplies. The quality, quantity and timing of water yield from mountain catchments are therefore of considerable social and economic importance.

Mountains remain the dominating control on the flows of most major rivers for considerable distances downstream. The Ganges, whose source resides in the Himalayas, provides water for domestic needs, irrigation, industrial use and urban supplies to a population of over 400 million in the Indian lowlands. Similarly, a small country like Switzerland is often referred to as 'Europe's water tower', because the Swiss Alps are the main source of water for the four largest western and central European rivers, namely the Rhine, the Rhône, the Danube (via its tributary the Inn

River) and the Po. The Alps occupy just over 10 per cent of the area of the Rhine's drainage basin, but contribute to over 50 per cent of its flow in summer. In arid and semi-arid regions, mountains often contribute over 90 per cent of river flows.

Water falling in a mountain river catchment, which is not stored (in the form of snow or in lakes) or evaporated, can take one of two paths, namely surface runoff or underground flow. In some instances, water can change from one path to the other. Flow above ground is more rapid and subject to greater fluctuations than that below ground. As it moves below the surface, some water is held in the soil and from there may be lost by evaporation from bare surfaces or transpiration from plants. Different kinds of vegetation transpire at different rates; under certain conditions of rainfall and evaporation, the type of vegetation can have a significant effect on water yield. Sediment loadings can be exceptionally high in many rivers originating in mountains, and reflect geomorphologic processes. Unless slowed down by major lakes within mountains themselves or in adjacent lowlands, or by engineering works designed to control torrent flows, rivers will carry their heavy loads unimpeded into the plains. This can be beneficial for agriculture because these sediments create rich and fertile soils, but excessive loads can have adverse consequences such as flooding of populated lowland areas.

Mountain hydrology has a number of particularities which determine the amount and timing of water discharge. There is very important natural storage in mountainous regions related to low-temperature effects (seasonal snow cover, glaciers and ice sheets), and to storage in karst-type (limestone) environments. However, if the water storage capacity of snow and glaciers is not available, the soil layer over the impervious bedrock is usually too thin to hold much moisture, and rainwater rapidly runs off, causing floods in locations downstream. Overland flow leading to soil erosion contributes to the degradation of watersheds by decreasing the already-small soil water-holding capacity. In regions affected by winter precipitation, temporary storage of water in the form of snow and ice provides delayed runoff. Snow cover at high elevations ultimately melts in spring and early summer, thereby contributing to a large discharge in many river basins. This is of major significance in some of the more elevated mountain chains of the world in terms of water availability in the source region of major rivers (Steinhauser, 1970). Hydrological systems are sensitive to a number of snow parameters, in particular snow cover duration, which has been shown to vary linearly with altitude and is also a function of slope orientation (Slayter *et al.*, 1984). Snow depth also varies with altitude, orientation and topography (Witmer *et al.*, 1986; Föhn, 1991). A further contributor is snow-melt runoff, which feeds into the hydrological system of mountains; this is determined by temperatures and surface energy balance during the spring season (Collins, 1989; Chen and Ohmura, 1990).

Artificial reservoirs in mountainous areas may be constructed for different purposes such as flood protection, water supply, recreation and water power. Vast amounts of potential energy are available in mountain rivers, which can be used for hydroelectric power generation. Numerous countries endowed with mountains and sufficient water supply, such as Norway, Switzerland or New Zealand, use this resource to satisfy a significant fraction of national energy demand.

Mountain regions are very susceptible to disasters in which water plays a key role. They are particularly flood-prone owing to the enhanced orographic precipitation, steep slopes with a high runoff coefficients, thin soils overlying impervious bedrock (which cannot buffer excess precipitation), and limited valley storage. In February 1995 the Rhine basin in Germany and the Netherlands was particularly affected by flooding, following torrential rains over much of Western Europe and early snow-pack melting in the Alps. Cities such as Cologne in Germany underwent extensive flooding, and dikes in the Netherlands threatened to yield to the exceptional flows.

In addition to floods caused by intensive rainfall and/or snowmelt, as well as events of rain falling on frozen surfaces, there are numerous examples of floods whose causes are not directly related to climatic extremes (Starosolszky and Melder, 1989). Glacier outbursts, avalanches, landslides, debris flows, ice jams, volcanic events and earthquakes may lead to the damming of a river, or reduce the effective volume of a reservoir or a lake, thereby leading to catastrophic flooding.

2.3 Mountain cryosphere

Elements of the cryosphere, namely snow, ice and permafrost, are located in most high mountain regions of the world; the height at which these features may be found is a function of altitude, latitude, and the degree of continentality (i.e. the availability of water which can fall out as snow). Cryospheric processes in high mountain ranges have largely contributed to the current morphology of these regions, through spectacular erosion processes. Because of its extreme sensitivity to climate, the mountain cryosphere is an excellent indicator of climatic change.

Snow cover and duration play a key role in a number of environmental and socio-economic systems in mountain regions. The behaviour of hydrological systems and mountain glaciers is closely linked to the timing and volume of snowfall and snow-melt (Barry, 1992a). In countries such as Switzerland, Norway or New Zealand, where between 60 and 90 per cent of electricity production is from hydro-power, energy supply is highly dependent on, and sensitive to, changes in snow amount and duration. Mountain ecosystems, particularly vegetation at high elevations or ecosystems which experience long winters, depend on snow cover to protect numerous dormant species from frost damage during the winter season. The economic value of snow in many mountain regions, especially in North America and the European Alpine countries, is paramount for winter tourism. Many small ski resorts in Switzerland, France and Austria have faced severe shortfalls in earnings whenever snow has been sparse or absent during major vacation periods, such as at Christmas and during the February school recess (Abegg and Froesch, 1994).

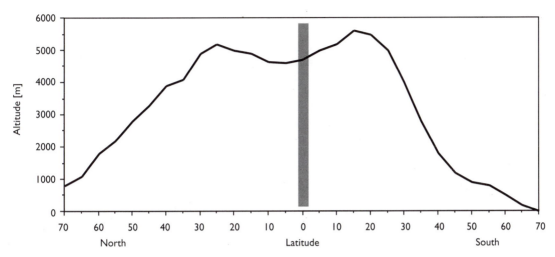

FIGURE 2.5 Average annual snow line height as a function of latitude

As an indicator of climatic change, snow is an interesting variable, because it is dependent not only on temperature but also on precipitation. Unlike most meteorological variables taken individually, records of snow depth, spatial extent and duration not only are a function of diurnal values of temperature and precipitation, but are based on the history of these variables over a period prior to the observation itself; these include ageing and packing of the snow over time. As a result, the interpretation of 'instantaneous' snow statistics is far from trivial because a given value of snow depth recorded on a particular day will generally have little relation to the temperature observed on that same day. Over longer periods, however, snow-pack records averaged over monthly or yearly periods provide a useful insight into interannual or longer time-scale climatic fluctuations, since the day-to-day precipitation events and temperature fluctuations are smoothed out, allowing the longer-term fluctuations of snow amount to be analysed.

The snow line is the boundary between the seasonal snow-pack which melts at the end of the winter season, and the permanent snow which remains year-round. Snow line is dependent on numerous factors other than simply climate, in particular local site characteristics such as slope orientation, and the type of vegetation cover. It is a difficult parameter to measure, because of its high temporal and spatial variability; however, when averaged over space and time, it does provide an approximate measure of the glacial zone and delineates the distribution of most mountain fauna and flora. Figure 2.5 illustrates the mean altitude of the snow line as a function of latitude (Charlesworth, 1957).

Regional snow lines are at sea level in boreal regions, and highest in the tropics; the most elevated snow lines are located in the arid Andes (Acatama, Chile) and the Tibetan Plateau, where they exceed 6000 m. In the equatorial zone, the depression of the snow line reflects the greater amounts of precipitation and cloudiness which generally occur close to the equator, whereas the reverse is true for the subtropical regions located between 20° and 30° latitudes N and S. The degree of continentality also has a marked influence on the snow line; the west-facing slopes of the New Zealand Alps of South Island receive heavy precipitation year-round, such that snow and ice come very close to sea level at a latitude of 45°S. At comparable latitudes north of the equator, the Rockies in Montana and Wyoming have a much higher snow line as a result of their distance from the rain-bearing winds of the Pacific Ocean. In addition, the coastal ranges of Oregon and Washington states intercept a significant fraction of the precipitation as it moves eastwards across North America.

A particularity of snow in mountains is the potential for avalanches. An avalanche is an abrupt release and downward movement of snow, triggered by an unstable snow-pack and/or rapidly changing temperatures or snow conditions. Avalanches are extremely destructive and take a high toll on human life in inhabited mountain regions. One potentially destructive form is the powder-snow avalanche, which occurs under cold conditions when snow is loose; the avalanche takes the form of a two-phase flow, with a mix of air and powder snow which has the characteristics of a fluid. Velocities of a mature powder avalanche can exceed 100 m/s; much of the destructive capacity is linked to the intense turbulence within the flow. Because the snow particles tend to settle through the effects of gravity, while the air pursues its downward movement, there can be instances of damage well beyond the normal avalanche zone. Slab avalanches are also extremely destructive, and occur when snow breaks away from part of the slope. According to its texture and degree of compactness, the avalanche can move downslope with sufficient internal cohesion for it to travel as a single heavy mass, until it disaggregates on contact with natural or artificial obstacles. Anything buried in the path of such avalanches will experience severe damage, and chances of survival for persons caught in the moving snow are remote, unless they are rescued within minutes of the event.

In populated mountain regions, and for communities where tourism in winter is a major resource, avalanche control techniques include the construction of avalanche fences, the afforestation of slopes, the channelling of valley sides to divert avalanches, and tunnels or concrete galleries over roads and railways. These have considerably reduced loss of life and damage to infrastructure, particularly in avalanche-prone countries such as Austria or Switzerland. In remote mountain regions, accumulation from avalanches can favour the accumulation of snow on glaciers, thus 'feeding' the glaciers below their normal accumulation zone.

Glaciers form as a result of the compaction of snow, which, under favourable climatic or topographic conditions, accumulates over a long periods and where there is an excess of solid (snow, ice) precipitation over melting. In some areas, however, where the temperature rarely rises as high as the melting point, accumulation of ice takes place through a recurrence of sublimation and recrystallization processes. Today's mountain glaciers are mere remnants of the last major ice age, which occurred some 15,000–20,000 years ago.

If a glacier is sufficiently thick, or the valley slope in which it forms is sufficiently steep, the ice will begin to flow under the influence of gravity and the pressure of the accumulated snow and ice layers. As it creeps downwards, the ice undergoes deformations typical of plastic materials. The speed of a glacier is determined by its plasticity, and the temperature at the base of the glacier. If this temperature is close to or above the freezing point (as in many mid-latitude glaciers today), then slippage of ice over the underlying bedrock will be enhanced by the lubricating effect of a film of water at the interface between ice and rock. In colder regimes, i.e. at high altitudes or high latitudes, the ice remains frozen to the bedrock, so that ice movement is slower and takes place through plastic deformation only.

A glacier perpetually moves downslope, but according to its mass balance, i.e. the equilibrium between accumulation and ablation, it can experience significant changes in length and volume. As the glacier flows down the valley to a lower altitude where it is not replenished by snowfall, it melts or wastes away, the melt-water forming the source of streams and rivers. If the accumulation rate is much lower than the rate of wasting, then the glacier enters into a phase of retreat. The equilibrium-level altitude (ELA) is defined as the height at which the accumulation by snowfall exactly balances the ablation by evaporation and melting; changes in the ELA are indicators of changing climatic conditions.

When a number of alpine glaciers flow together in the valley at the foot of a range of mountains, they frequently form glacier sheets known as piedmont glaciers. Glaciers of this type are especially common in Alaska, the largest of which is the Malaspina Glacier, with an area close to $3900\,km^2$. Icecap glaciers are common in boreal climates, and can be found in regions such as Baffin and Ellesmere Islands (Canada), and to a lesser extent in Svalbard in the Arctic Ocean. At very high-elevation sites, icecap-type glaciers can be found; these are generally located on rounded summit or crest surfaces in regions such as the southern Andes of Argentina and Chile, and in the northern Rockies of British Columbia and Alaska. On a much smaller scale, icecaps are also found in the European Alps (Claridenfirn, central Swiss Alps; the Breithorn Plateau and the Theodul Glacier on the Italian–Swiss border close to Zermatt, for example).

As a glacier moves down a valley, it erodes the landscape in a manner very different from fluvial erosion; glaciated mountains are rugged and typified by pyramidal peaks, jagged ridges and vertical-sided valleys. Rocks in the path of a glacier are ploughed out of the way, and rocks beneath it are broken up by frost action and then carried away. The rocks embedded in the bottom of the glacier act as abrasive particles, scouring the surface beneath. Glacial erosion leads to characteristic U-shaped valleys, and combined ice and frost action result in the formation of cirques and distinctive sharp peaks, some of which have become major attractions for tourists and mountaineers (e.g. the Matterhorn near Zermatt, Switzerland).

Mountain permafrost is important in terms of the cohesiveness of soils and rocks, and therefore the stability of mountain slopes. Permafrost occurs in regions of low temperatures and low to moderate liquid precipitation (less than $1000\,mm$ annually); it consists of a mixture of ice, soil and bedrock and can penetrate down to depths of over $100\,m$ in places. The bulk of permafrost occurs outside mountain regions (French, 1996) and is present in about 25 per cent of the surface area of continents (100 per cent in Antarctica; 80 per cent in Alaska; 50 per cent in the former Soviet Union; 45 per cent in Canada). It is estimated that about 1 per cent of the global supply of fresh water is stored within the permafrost.

Because of the insulating capacity of most soils, permafrost can subsist at certain depths below the surface even during the summer period when snow has melted at the surface. There are different forms of permafrost and manifestations of its presence in mountain areas. The active layer represents the first 2–5 m below the surface where annual temperature cycles lead to summer melting; the permafrost proper occurs below the region of seasonal thaw. In many mountain regions, so-called rock-glaciers are formed when surface material undergoes deformation by movements within the permafrost layers (Haeberli, 1985). According to the exposure, the length of the snow season, atmospheric dynamics and the nature of the surface (e.g. compact soils or large boulders, which conduct heat into the soil differently), mountain permafrost can be continuous (particularly at high elevations with a short summer season) or discontinuous (at medium elevations), or occur in isolated patches which have microclimates favourable for sustaining permafrost.

In Switzerland, about 5 per cent of the total surface area of the country is affected by permafrost (roughly $2000\,km^2$), which is more than the area covered by its glaciers ($1300\,km^2$, or 3 per cent of the surface area of Switzerland). Surfaces recently evacuated by retreating glaciers undergo the formation of new zones of permafrost, because the surface is in contact with the atmosphere rather than the ice, which tends to insulate the ground from frost and persistently low temperatures.

The melting of the ice within the permafrost layers reduces the stability of most slopes, and debris flows, mudslides and rockfalls can result. Infrastructure at high elevations can come under threat in regions of marginal permafrost, particularly hydro-power dams, cable-car pylons and anchor points for various constructions.

2.4 Soils

Soils are the result of climate and living organisms acting on parent geological material, with topography or local relief exerting a modifying influence; furthermore, time is an essential factor in the soil-forming process (Jenny, 1941). Under similar environments in different places, soils are similar.

A soil typically consists of distinct layers called horizons, which make up a soil profile. Each layer has its own distinguishing features, namely colour, texture, organic matter content, and acidity or alkalinity. Water in soils tends to percolate downwards, and transports material from the layers close to the ground surface to lower layers. Well-defined horizons are the result of undisturbed, hence long-term, soil processes. Mountain soils rarely contain such well-defined soil layers, because of the extremes of climate, rapid erosion processes, high acidity and low fertility. Mountain soils consist of heterogeneous and discontinuous regions which are constantly undergoing changing site conditions (Price, 1981).

The effects of topography or local relief, parent material and time on soil become apparent particularly at the local and regional scales. In humid regions, wet soils and the properties associated with moisture tend to occur in flat, low-lying regions, while better-drained soils generally form in regions where topography is important. In arid regions, the differences associated with relief are associated mainly with salinity. In a local environment, different soils are associated with contrasting parent materials, such as residuum from shale and from sandstone.

Climate has a dominating influence on soil properties. Rainfall and temperature control the intensity of leaching and the weathering of soil minerals, thus having a major influence on the chemical properties of soils, such as acidity, alkalinity and salinity. Acidity is associated with leached soils, whereas alkalinity occurs predominantly in drier regions. Within a given region, the acidity or alkalinity level of a soil depends on acid inputs from vegetation, the microbial biomass and the capacity of the minerals which make up the parent material to resist the acidifying effects of leaching.

Vegetation, itself a function of climate in mountains, gives the soil its distinctive character, in that it controls the quantity and variety of organic material which is recycled into the soil. The proportions of vegetation cover and type are essential in determining thermal properties of the soil and its resistance to erosion. Because of the reduction of biological activity with height in mountains, soil-forming processes also weaken, so that at high elevations, soil formation is extremely modest; soils in high mountain areas are shallow, contain limited amounts of nutrients and are therefore not capable of sustaining much vegetation. Soils can change abruptly on mountain slopes as a function of sharp vegetation transitions; one of the most prominent soil boundaries in mountains and uplands is to be found close to the timber line, where conifers yield to alpine meadows and soil acidity changes over short distances. There are many other interactions between mountain plants and soils. Perhaps the most important one is the mechanical stabilizing effect by plants; any change in the plant cover will affect soil stability. When slow-growing species with their extensive root systems and below-ground stems are replaced by fast-growing species, which invest more in above-ground structures, soil stability on slopes will decline (Körner, 1994).

Mountain slope characteristics have an important influence on the amount and rate of runoff, as well as on sedimentation associated with runoff (e.g. Parsons, 1978). Slope length has considerable control over river drainage and potential accelerated water erosion. Slope aspect is the direction towards which the surface of the soil faces, and it plays a key role in determining soil temperature and moisture (Ollier, 1976).

Direct human interference can lead to modification of soil characteristics; irrigation, liming or planting of exogenous plant or tree varieties may upset the natural acidity or alkalinity of soils. Atmospheric pollution, in particular those trace elements such as sulphur dioxide related to 'acid rain', is also a major anthropogenic cause in the modification of soil characteristics, both in mountains and in lowland regions.

Erosion is the detachment and movement of soil material. The process may be natural or accelerated by human activity. Local land-scape, land cover and climatic conditions determine the rate of erosion. Natural erosion has sculptured landforms on the uplands and built landforms on the lowlands. Its rate and distribution in time controls the age of land surfaces and many of its soil properties.

Landscapes and their soils are evaluated from the perspective of their natural erosion history. Buried soils, stone lines, deposits of wind-blown material, and other evidence that material has been moved and redeposited are helpful in understanding natural erosion history. Thick weathered zones that developed under earlier climatic conditions may have been exposed to become the material in which new soils formed. In landscapes of the most recently glaciated areas, the conse-quences of natural erosion, or lack of it, are less obvious than where the surface and the landscape are much older in geological terms. Even in recently glaciated regions, post-glacial wind, water or chemical erosion may have redistributed soil materials on the local land-scape. Natural erosion is an important process that affects soil formation; indeed, erosion is capable of removing soils formed in the natural landscape. Accelerated erosion is largely the consequence of human interference with the natural environment. Primary causes are land-use changes associated with agriculture and pastoral practices, exploitation of timber, channelling of hydrological systems and, in some regions, urbanization and tourist infra-structure.

Classification of mountain soils is difficult because of the diversity of the mountains themselves and the complex set of interactions which are possible between local climates, vegetation distribution, and geological and orographic features. Soil types can vary from one slope to another as a function of exposure and steepness; soil processes and characteris-tics observed at a particular altitudinal level in a mid-latitude mountain belt are unlikely to be found at the same level in a tropical mountain region.

Price (1981) identifies three major regions where a number of mountain soil types are likely to be present, namely:

- humid mid-latitudes
- arid regions
- humid tropical areas.

Each broad region contains specific soil types which reflect the climatic and ecological condi-tions. In mid- and high latitudes, temperature controls the intensity of soil formation, while in more arid regions, the altitudinal distribu-tion of precipitation is the dominant climatic factor (Messerli, 1973; Hanawalt and Whit-taker, 1976). In the tropical mountains, both temperature and moisture are important; one of the most important distinguishing charac-teristics of tropical mountain soils is lateriza-tion, which consists in the selective leaching of silica from the upper soil layers while leav-ing iron and aluminium oxides at the surface (Grubb and Tanner, 1976; Edwards and Grubb, 1988). Lateritic soil formation is depen-dent on the high temperatures and abundant moisture common in many tropical regions; the unusual red colour of these soils is due to the high iron and aluminium content which the leaching has left behind.

2.5 Ecological systems and biodiversity

Because of their great altitudinal range, moun-tains such as the Himalayas, the Rockies, the Andes and the Alps exhibit, within short hori-zontal distances, climatic regimes which are similar to those of widely separated latitudinal belts; they consequently feature high biodiver-sity. Although the absolute number of species is generally smaller than in lowland areas, mountains feature unique ecosystems which have survived human encroachment because of the fact that mountain environments are far more difficult to exploit than plains. On a global scale, plant life at high elevations is primarily constrained by direct and indirect effects of low temperature and, in addition, to reduced partial pressure of carbon dioxide. Other atmospheric influences, such as increased radiation, high wind speeds or insuf-ficient water supply, may come into play on a regional scale, but exert no globally uniform characteristics of high mountain systems (Lauscher, 1977; Körner and Larcher, 1988). The parallel between latitude and altitude is therefore not always valid, because conditions in analogous zones may be very different. The very local site characteristics of regions of complex orography, reflected in differential absorption of radiation according to slope aspect and exposure, and the distribution of precipitation (rain shadow or enhanced lee precipitation) have no latitudinal counterparts. This is to a large extent reflected in the physiological aspects of certain plants.

Altitudinal vegetation distribution has traditionally been used to characterize mountain environments. This distribution is strongly influenced by climatic parameters, but not exclusively. Topography, edaphic factors and, in many areas, human activities and the disturbances they generate strongly interfere with the potential distributions suggested by large-scale climatic parameters alone. The effects of climatic variables, especially temperature and precipitation, on the distribution of plant life has long been recognized. Humboldt (1817) was among the first to describe the relationship between climate and vegetation. Various altitudinal zonation classification schemes have been developed to reflect site-specific conditions of

climate and vegetation in mountain areas around the world. For example, the Köppen (1931) climate classification system is based on major natural vegetation patterns. Holdridge (1947) devised a life zone classification more appropriate to the complexities of tropical vegetation. Holdridge's zones are delineated according to bio-temperature, which refers to all temperatures above freezing, with all temperatures below freezing adjusted to 0°C. The assumption was that, from the perspective of plant physiology, there is no real difference between 0°C and temperatures less than zero, since plants are dormant. The life zones are thus defined first according to a climatic variable, i.e. degrees mean annual plant temperature, and not according to degrees latitude or metres of elevation. The broad, climatically defined life zones are further subdivided into associations on the basis of local environmental conditions and actual vegetation cover or land use.

The zonation patterns of vegetation differ according to the location of the mountains along a latitudinal gradient, as shown in Figure 2.6 (Beniston *et al.*, 1996). The basic zones of vegetation encountered when going upwards on a mountain slope include submontane, montane, subalpine, alpine and

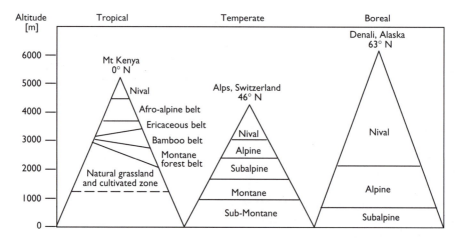

FIGURE 2.6 Schematic diagram of vegetation zonation in tropical, temperate and boreal mountain regions. In the tropical zone shown in this example, some belts are tilted because of the large differences in precipitation which occur on one side of the mountains with respect to the other

nival. Mountains in temperate and boreal zones are characterized by seasonal climates with a well-defined growing season, whereby plants can exploit the short, warm summers. Growing season is defined in terms of snow cover and duration on the one hand, and temperature thresholds for a particular plant, on the other. Many plant species at high altitudes are capable of surviving harsh conditions as a result of the onset of a hardening process. Amount and duration of snow cover influence high mountain vegetation by determining growing season and moisture conditions. In the alpine zones, exposed, windy ridges are generally covered with xeric dwarf shrubs and lichens. Depending on nutrition and soil moisture, more mesic dwarf shrub heaths and mesic low herb meadows are found on lee slopes of intermediate snow cover. The late-exposed snow patches have hygrophilous herb communities. Vascular plants may be completely lacking owing to the very short growing season. Mountain plants grow more slowly but have photosynthetic capacity equal to or greater than that of lowland plants (Körner and Diemer, 1987; Körner and Pelaez Menendez-Riedl, 1989). It has been shown that this has partly to do with the fact that most alpine plants produce only one leaf generation (Diemer *et al.*, 1992), and have somewhat greater below-ground carbon investments (Körner and Renhardt, 1987). The longer the period during which photosynthesis is halted owing to snow cover, the greater the difficulty certain mountain plants will have in achieving a long-term positive carbon balance.

Vegetation zonation on mountains in the tropics does not replicate the latitudinal belts of vegetation of the middle and higher latitudes, even though mountain peaks may extend well above the snow line. In part this is due to the fact that the seasonal cool and cold temperatures of the middle and high latitudes are not experienced in the tropics. Instead, it is diurnal temperature patterns that are important, while seasons are distinguished on the basis of precipitation patterns. While there are similarities in vegetation types in the mountains of tropical Latin America and East Africa, there are also distinct differences which reflect the very different geological history of the two continents. The Andean Cordillera of South America is a continuous range of mountains stretching along the entire west coast of the continent, the result of mountain building associated with continental drift. In East Africa, a discontinuous chain of mountains ranges from Ethiopia to South Africa. Individual peaks and ranges have different ages, origins and geologies. Instead of forming a long corridor, as in the case of the Andes, Africa's mountains resemble islands rising above the surrounding plains (Hedberg, 1964); Box 2.2 provides an overview of the vegetation zonation of mountains such as Mt Kenya or Mt Kilimanjaro in East Africa.

A unique feature of tropical mountains is cloud forests, which exhibit a high degree of biodiversity, such as in parts of the Central American cordilleras, the tropical Andes, the highlands of Papua New Guinea and exposed slopes in the Kenyan mountains.

Plants in the alpine zone of high mountain systems are not particularly limited by water supply in most parts of the world, since precipitation tends to increase with height. There are certain notable exceptions, such as the continental mountain regions of Central Asia (Izmailova, 1977), rain-shadow areas (Richard, 1985), exposed ridges (Oberbauer and Billings, 1981) and slopes such as the Andes (Gonzalez, 1985). Even hot, exposed scree slopes, often supposed to be dry habitats, have been shown to be places of particularly good water supply (Pisek and Cartellieri, 1941). Most alpine soils under closed vegetation are deeper and the roots penetrate further into the layers of weathered substrate. Reduced water supply to leaves rarely poses a problem to alpine plants, but may locally or periodically reduce their activity.

A special form of water stress may occur in evergreen species in later winter, when soils may still be frozen, while evaporative driving forces are already strong. Winter desiccation of leaves may lead to shoot damage in certain isolated conifer species (Baig and Tranquillini, 1980), but this takes place only in certain regions and cannot be considered to be a general problem for mountain plants. The

Box 2.2 Ecoclimatic zones in East African mountains

On Mt Kenya and some of the other high summits of East Africa, several ecoclimatic zones can be identified (Njiro, 1998), namely:

- nival zone, above 4300 m
- alpine zone, above 3500 m
- heath zone, above 3000 m
- hagenia/hypericum zone, above 2800 m
- bamboo zone, above 2400 m
- forest and savanna grasslands subsist below about 2400 m.

Plants are adapted to high radiation levels and low temperatures; at the higher elevations, ultraviolet (UV) radiation can be strong, particularly on cloud-free days. The exposure to high UV intensity can have adverse effects on some of the life forms which exist on Mt Kenya. Some plants have morphological traits that protect them from strong radiation through reflection. These include the thick and light-coloured indumentum (dense hair) covering one or both sides of the leaves, and/or bracts, peduncles, and stems of plants such as *Senecio brassica*, *Hiechrysum newii* among others. These are more common in the alpine belt than in the lower parts of the mountain.

The East African montane ecosystems respond to high rainfall, which occurs during the two peak precipitation periods of April–May and November–December. Diurnal cycles represent the significant temperature variation in the Afro-alpine zone. During the day, high temperatures occur as a result of direct solar radiation through a clear atmosphere. During the night there is a strong loss of heat through long-wave radiation, which frequently results in overnight frost (Hedberg, 1995). This phenomenon has marked implication on species adaptations in the zone. Protection of plants from freezing is achieved through formation of tussocks and rosettes, among other developments.

In the lower montane forests (between 2100–2400 m), the tree species include camphor (*Ocotea usmbarensis*), cedar (*Juniperus procera*), olive (*Olea africana*) and *Podocarpus* spp. The bamboo zone (2400–2800 m) is the home to the giant grass bamboo (*Arudinalia alpina*), which grows as a thick forest attaining heights of 12–15 m. Interspersed are numerous flowering herbs. The hagenia/hypericum zone (2850–3000 m) has forests with patches of bamboo thickets and open glades. The heath zone (3000–3500 m) consists of open moorland with smooth tussock (*Deschampisia flexiosa*) and tussock grasses (*Eleusina jaegri*) interspersed with other plants varying according to altitude. The alpine zone, located between 3500 and 4350 m, contains lobelia species, which are found in groups along wet valley bottoms. The most conspicuous inhabitant of this zone is the rock hyrax. The nival zone, above 4300 m, comprises rock and ice, and the most common life form consists of a variety of colourful lichens.

relatively balanced water status in mountain plants is emphasized *inter alia* by Körner and Cochrane (1983) and Körner *et al.*, (1986).

Tree lines represent a major ecological discontinuity as trees reach their limit of climatic tolerance; on mountains, this occurs at the interface between subalpine vegetation and low-growing alpine vegetation. There is much debate on the causes of tree lines, but there is consensus that climate is one of the major controlling factors on the carbon balance and mortality of tree species (Slayter and Noble, 1992). The subalpine tree line is a typical ecotone which is controlled by carbon balance, where the life form with the smallest photosynthetically active biomass fraction (leaf mass ratio only 1 to 4 per cent of total biomass) is replaced by life forms with greater leaf mass ratio such as shrubs (10–15 per cent) and forbs (20–25 per cent; Körner, 1993). This phenomenon is observed at all latitudes. Interestingly, temperate and boreal zone tree lines are situated at similar isotherms of the warmest month (generally considered to be 10°C) as are subtropical and tropical tree lines, despite the much shorter growing season of temperate and high-latitude trees (Wardle, 1974). As the tree limit is approached, factors other than competition for solar energy to maintain a positive carbon balance assume greater significance; the response of trees is thus to reduce the amount of carbon allocated to wood production and to reduce growth (Huggett, 1995). Photosynthesis at high elevations is thus one of the keys to explain the rapid transition of ecosystems at the tree line (Tranquillini, 1979; Smith and Knapp, 1990).

In some mountain regions, non-climatic factors influence the tree line; in Chile, the altitude of forests, as well as forest structure and composition, are linked to volcanic activity (Veblen *et al.*, 1977), while biotic influences appear to determine tree lines in the Caucasus (Armand, 1992). In a more recent focus on the mechanisms behind tree-line ecotones, Körner (1998) notes that the often-cited 10°C isotherm of the warmest month is an empirical relationship which holds for North American and European mountains, but less well elsewhere. He suggests that below-ground processes are crucial and that mechanisms associated with photo-assimilation *per se* have less predictive value than previously believed. The presence of numerous trees at their upper limit of growth may, through shading effects of the branches and tree-trunks, cut off the energy required by the root system. The trees therefore contribute to their own upper limits; such a hypothesis also helps to explain why isolated trees, whose roots are not shaded by neighbouring trees, can often be found above the main tree line, and why the tree line itself is not always a sharp, step-like discontinuity.

2.6 Human environments

Human populations have lived in most mountain regions for centuries and even millennia. In certain regions of the world, however, notably those settled by Europeans who displaced indigenous populations, the history of today's predominant population dates back only to recent centuries. Grötzbach (1988) has described these two broad categories as, respectively, 'old' and 'young' mountains. The latter, in North America, Australia and New Zealand, are further characterized by sparse settlement, extensive market-oriented pastoral agriculture and forestry, and, in the post-Second World War period, the rise of tourism as a major economic force (Price, 1994).

'Old' regions, particularly the European mountains, most of which are relatively densely settled, can be further divided into three broad categories. The first of these is characterized by the decline of traditional agriculture and forestry, which subsist essentially through government subsidies. During the post-war period, tourism has grown rapidly, and is in many regions today one of the principal income-earners for a country as a whole (e.g. Austria), and the mountain communities in particular. The most recent stage of development is characterized by the globalization of the markets. Under the mounting pressure exerted on the agricultural sector to deregulate, even the highly protected and subsidized mountain agriculture in Switzerland or Austria is today encountering problems. Nowadays, many problems specific to the Alps, such as transport and transit problems, demand solutions that can be found only within the overall European context; policies relevant to the European Alpine region are no longer determined exclusively within the Alps themselves, but increasingly in Brussels, London or Paris. Today, it is not so much *where* the decisions are made that is important, but *who* the political and economic actors are that are taking the decisions.

While in some regions, such as Western Europe and Japan, mountains are experiencing depopulation (Yoshino *et al.*, 1980; Price, 1994), in general, land-use pressures are increasing because of competition between refuge use, mineral extraction and processing, recreation development, and market-oriented agriculture, forestry and livestock grazing (Ives and Messerli, 1989; Messerli, 1989). Areas without significant tourism have in the past 30–40 years experienced extensive depopulation, as in parts of the French Alps or the Pyrenees. The tourism sector, which used to be an equalizing factor, has declined dramatically in the Alps in the 1990s, in part due to competition from far-away destinations which are today accessible from Western Europe and elsewhere at relatively low prices. Changes in the tourism sector are also felt in related industries, particularly construction and services (Pils *et al.*, 1998). The socio-cultural responses to rapid socio-economic change are highly complex. In the economically developed Alpine regions, there is a dichotomy between the modernization of the mountain communities and the desire of most tourists to have access to traditional customs and architecture. However, it is impossible to transform an area and its population into a museum. The adverse effects of tourism are increasingly raising awareness of local populations to development and environmental issues, in particular the impacts of transit and tourist traffic, related air pollution, waste and water resources.

The second category of 'old' mountain regions includes most of the mountains of developing nations. Traditional subsistence agriculture is one of the main economic activities in the upland areas of countries such as Nepal, Bhutan, New Guinea and Peru. The inhabitants of upland zones of many developing countries are essentially mountain peasants who practise some form of transhumance, in particular in the Andes, the Himalaya and sub-Sahelian Africa. In other parts of the developing world, nomadic agriculture is experienced even today, although there is a clear tendency towards sedentary lifestyles; nomadic practices are observed in the High Atlas of Morocco, parts of the Middle East, and in the uplands of Pakistan, for example. In all these regions, tourism is becoming economically important in small, well-defined areas, often with considerable effects up to the national scale (Price, 1993). In some countries, such as Nepal, Kenya or some of the Andean states, tourism is the primary source of foreign exchange because of the attractivity of the high and remote mountains for hikers and climbers, among other factors. In many countries, the development of tourism is today impaired by political instability or warfare (e.g. Afghanistan, Tajikistan, Ethiopia, Peru).

Technical development inputs have an important contribution to make to sustainable development in highland areas. These can, however, also exacerbate problems if introduced without taking into account local traditions, bodies of knowledge possessed by indigenous people and the linkages between agriculture, animal husbandry, forestry and other subsistence practices. Similarly, interventions based on models of development in lowland areas are often unsuitable for fragile mountain systems; these can result in highly unbalanced terms of trade that characterize highland communities' interactions with the more dominant lowland political economies (Mehta, 1995).

The third category of 'old' mountains and uplands was, until the early 1990s, characterized by planned economies in the agriculture, forestry and mining sectors. These include the mountains of the former Soviet Union and the Carpathians in Romania, Slovakia and Poland. With the fall of the communist regimes in all these countries, the social, economic and political structures of these regions are now in a state of flux. In parts of the Carpathians and the Caucasus, tourism based on the capitalist model is gradually replacing the former structure in which the supply of 'tourists' was guaranteed. China still maintains a centralized, planned economy, although the mountains, being far from centres of power, do not experience the degree of collectivization or nationalization common in the more productive lowland regions. Tourism is developing in isolated regions in south-western China, often with little of the infrastructure which the 'Western tourist' is seeking. In addition, the political problems in Tibet act as a deterrent to any extensive increases in tourism in this heavily controlled and militarized part of the country.

Mountains in developing nations face stress from increasing human population. In many of these regions, pressures exist to develop market-based agricultural systems, changing from a largely local economy to a national

Box 2.3 Technological upheavals and consequences for traditional lifestyles in Indian upland regions

Subsistence agriculture, which allowed indigenous populations to survive in mountains and uplands in India, is gradually giving way to modern technologies, which are not always adapted to local conditions (Mehta, 1995). The loss of resilient traditional varieties of crops tends to affect subsistence activities. For example, old species of millet and wheat cultivated throughout the Himalayas are highly adapted to the fragile conditions of highland cultivation. They are, in addition, nutritious food sources for both humans and animals. In many areas, however, changing dietary habits (particularly among the young) favouring wheat and rice, along with the greater use of hybrid wheat varieties, has led to the devaluation of millets, which are often cultivated on marginal-quality lands and in smaller quantities than in the past. This situation has also affected animal husbandry strategies. The hybrid seeds produce shorter and thicker stalks, which withstand high winds and heavy rains; nevertheless, they are less resistant to fluctuations in rainfall and temperature, and are not favoured by peasants as animal feed. As a consequence, in many areas peasants now rely more heavily on forest leaves and grasses, a trend which ultimately has important implications for the well-being of local forest resources.

Another cost of new technologies is the emerging tension between indigenous folk systems of knowledge and exogenous scientific knowledge bases. Traditional subsistence methods are based on knowledge that has evolved through trial and error over centuries; these are highly adaptive to the constraints of specific highland microniches and are sustainable without long-term damage to the land (Shiva, 1988). In addition, these methods are not dependent on alternative market-based resources. The erosion of local knowledge affects the ability of households to adjust to emergencies and, in many instances, also leads to the devaluation of women, who are the main repositories of this knowledge. Furthermore, even if technological advances are to be a viable alternative to traditional subsistence agriculture, the income effects of shifts to cash cropping are highly dependent on existing transportation and storage infrastructure, credit facilities and pricing policies for the produce. In many instances, short-term gains for households are superseded by high levels of indebtedness and/or declining market prices for the crops (DeWalt, 1993).

and international one. In Africa, most of the mid-elevation ranges, plateaux and the slopes of high mountains are under considerable pressure from commercial and subsistence farming activities (Rongers, 1993). In unprotected areas, mountain forests are cleared for cultivation of high-altitude-adapted cash crops like tea, pyrethrum and coffee. Population growth rates in East Africa, among the highest in the world (Goliber, 1985), are concentrated in agriculturally fertile and productive mountain districts (Government of Tanzania, 1979). This is quite evident near Mt Kenya and Mt Kilimanjaro, and in the Usambara Mountains (Lundgren, 1980). Tourism is becoming economically important in small, well-defined portions of these regions (Price, 1993).

Particularly in tropical and semi-arid climates, mountain areas are usually wet, cool and hospitable for human dwelling and commercial cultivation. Human encroachment on mountain regions has reduced vegetation cover, thus increasing evaporation of soil-moisture, erosion and siltation, thereby adversely affecting water quality and other resources.

Anthropogenic stresses on mountain regions increased significantly in the second half of the twentieth century. Rapid technological developments and falling costs of transportation have increased human mobility into what used to be remote areas with difficult access. In the more developed countries, signs of encroachment into the mountain environments include industrialization of mountain valleys, communications infrastructure, and buildings for recreation and tourism. In the past, human interference with montane environments was essentially linked to agricultural practices; in regions with increasing populations and demands on food and resources, overgrazing and deforestation have transformed mountain environments. Some of the most overgrazed regions over the past two thousand years have been the eastern Mediterranean mountains, where pastoral practices have transformed once-forested mountains into arid and semi-arid regions.

Despite its obvious linkages to environmental degradation and poverty, health remains a much neglected issue within the broader field of mountain development concerns. The decline of forest, agricultural and water resources, intensification of work burdens, and the cold stress associated with living in the high mountains, and limited access to decent health care, are only some of the pressures that highland communities have to face in developing countries (Mehta, 1995). Most mountain communities lack adequate water supplies and proper sanitation facilities. In a number of areas, reduced access to fuel-wood is forcing communities to make adjustments to their diets, shifting from nutritious whole grains and legumes to less nutritious foods that require less cooking time (Dankelman and Davidson, 1991). As possibilities of cooking meals diminish, for example in parts of the central highlands of Africa, so does the sensitivity of the affected populations to disease increase.

2.7 Data for research on mountain environments

2.7.1 Climatic data

The European Alps are by far the best-known mountain area of the world in terms of weather and climate and related environmental characteristics (Barry, 1994). Several summit observatories have records spanning 100 years and there is a dense network of regular observing stations and precipitation gauges. Detailed climatic atlases exist for the Tyrol (Fliri, 1982) and the Swiss Alps (Kirchofer, 1982). The most intensive permanent observation sites, including high-elevation stations, are to be found in Switzerland and Austria. Through specific academic research initiatives, extensive studies have been made in the mountains of Scandinavia and the Carpathians of Central and Eastern Europe (Obrebska-Starkel, 1990).

The island of Hawaii is well known climatically through observations made at Mauna Loa Observatory since 1957, and various research programmes conducted on the windward and leeward slopes of the mountains in Hawaii and other islands. The climatic features of the Rocky Mountains, the Sierra Nevada and the Cascade Ranges in North America have been documented primarily through specialized university and other agency research programmes. The station networks are moderately good but permanent mountain observatories are lacking (Barry, 1994).

A moderate level of information on the climate of very high mountains is available in the case of the northern Andes (Monasterio, 1980) and the Himalayas of Nepal (Chalise, 1994). The former area has a reasonable station network, including some high-altitude stations. The large-scale climatic controls of this area are still poorly defined. In the case of Nepal, most stations are in the Himalayan valleys, but short-term measurements, particularly by recent Japanese glaciological expeditions, have provided additional information (Barry, 1992b). For the adjacent Tibetan (Qinghai Xizang) Plateau and Tian Shan there is now a growing meteorological literature.

Mountains where there have been discontinuous research programmes and only a sparse station network through which data are available include such diverse areas as New Guinea, Ethiopia (Hurni and Stähli, 1982), Mt Kenya, the Hoggar in Algeria, and the mountains of the Yukon and Alaska. In New Guinea, where there is only one permanent station above 2000 m, information on mountain environments has focused on ecological conditions in the Bismarck Range and on the glaciers of Mt Carstenz in Irian Jaya (Hope *et al.*, 1976).

2.7.2 Hydrological data

Klemes (1990) and Rodda (1994) have noted that there is a considerable lack of knowledge on mountain hydrology, despite the fact that the bulk of the world's fresh-water resources originate in mountains. There are a number of reasons which explain this paradox: difficult access to numerous drainage basins, sparse settlement with limited services, in particular electricity and telecommunications, and the harsh environments which hinder the development of hydrological observations in many mountainous regions. These conditions combine to make it difficult to install and to maintain instruments, which often require a special heavy-duty design capable of withstanding extreme environmental conditions. One of the characteristics of mountain areas, which adds to difficulties in the analysis of available data, is the large heterogeneity of environments and the great spatial and temporal variability of hydrological signals. Observation networks are nonetheless essential for the forecasting of precipitation and discharge, which is of primary importance not only for the mountains themselves, but also for regions located downstream that can experience flooding during abrupt increases in discharge.

2.7.3 Cryospheric data

Worldwide collection of standardized observations on changes in mass, volume, area and length of glaciers with time (glacier fluctuations), as well as statistical information on the distribution and characteristics of perennial surface ice in space (glacier inventories), is coordinated by the World Glacier Monitoring Service (WGMS) under the auspices of the International Commission on Snow and Ice (ICSI/IAHS), the Federation of Astronomical and Geophysical Data Analysis Services (FAGS/ICSU), the Global Terrestrial

Observation System (GTOS of WMO/UNESCO/UNEP/ICSU; WMO *et al.*, 1995) and the Division of Water Sciences of UNESCO. Data are periodically published (e.g. IAHS(ICSI)/UNEP/UNESCO, 1989, 1993, 1994; see also UNEP, 1992; Haeberli, 1995), and are incorporated in the World Data Center for Glaciology (WDC-A) and the Global Resources Information Database (GRID of GEMS/UNEP).

An extensive database of topographic glacier parameters is being built up in the context of regional glacier inventories. *The World Glacier Inventory – Status 1988*, published in 1989, is a guide to the existing statistical database on the worldwide distribution and morphological characteristics of glaciers as documented in regional inventories (some which are detailed, while others are preliminary; IAHS(ICSI)/UNEP/UNESCO, 1989). Repetition of such glacier inventory work is planned at time intervals which are comparable to characteristic dynamic response times of mountain glaciers (i.e. a few decades). This should help with analysing changes at a regional scale, and with assessing the representativeness of continuous measurements which can be carried out only on a few selected glaciers (Beniston *et al.*, 1998). In addition, glacier inventory data also serve as a statistical basis for extrapolating the results of observations or model calculations concerning individual glaciers (Oerlemans, 1993) and to simulate regional aspects of the effects of past and potential future climate change. The Global Change Working Group of the International Permafrost Association (IPA) has surveyed world researchers for sites that are monitoring ongoing change in permafrost character. Sites with measurements going back several decades, transects crossing the discontinuous to continuous permafrost transition, and intensely instrumented sites studying near-surface heat transport process have all been identified.

2.7.4 Ecosystem data

There is to date no comprehensive database for accessing ecological information specific to mountain ecosystems. Much research has been conducted on an individual basis and in concerted international actions (e.g. the UNESCO Man and Biosphere Project). Today, a number of recent research initiatives, particularly those coordinated by the International Geosphere–Biosphere Program (IGBP), are improving the situation regarding ecological information, particularly in remote regions of the globe, through transect studies in which vegetation and other parameters of environmental relevance are collected. One major initiative launched in the 1990s is the GTOS (Global Terrestrial Observing System; Heal *et al.*, 1993).

The Colorado Rocky Mountains and Scandinavian Mountains have also been sites of ecological research under the International Biological Program's Tundra Biome studies in the 1970s, and in the former area, related work has continued through Long-Term Ecological Research Program activities (Barry, 1986). European initiatives on long-term forest monitoring have been launched; they include the monitoring of forest sites in the Alps (Innes, 1994). In the Andes, microclimatic studies related to ecological conditions have

provided valuable supplementary information on ecology and links to other environmental systems (Monasterio, 1980). The impacts of climatic change on mountain ecosystems and nature reserves have become an increasingly important issue in the study of long-term biodiversity management and protection (Peters and Darling, 1985; McNeely, 1990; OTA, 1993; Halpin, 1994). Changes in global temperatures and local precipitation patterns could significantly alter the altitudinal ranges of important species within existing mountain belts and create additional environmental stresses on already fragile mountain ecosystems (Guisan *et al.*, 1995a), so that ecological information is of prime importance in understanding the functioning and feedbacks between vegetation and other systems, and to project potential modifications to vegetation zonation as a function of environmental change scenarios.

In many instances, palaeoecological data have provided much valuable information on mountain ecosystems and their evolution as a function of diverse stress factors. This aspect will be treated separately in Chapter 3.

Past environmental change in mountains and uplands

3.0 Chapter summary

Knowledge of past environmental change often provides valuable information on fluctuations in environmental parameters in the absence of significant anthropogenic forcing. Palaeoenvironmental studies can yield baseline conditions from which it is possible to deduce the importance of direct human interference compared with natural environmental changes. This chapter gives a brief overview of methods for reconstructing environmental conditions in the past, and outlines the status of current knowledge of changes in the physical and biological characteristics of mountains for periods ranging from several million years up till today. It will be seen that, as one approaches the twentieth century, the reliability of palaeoenvironmental information, in both temporal and spatial terms, increases dramatically.

3.1 Introduction

The evolution of mountains and their particular environmental characteristics are determined over the very long term (millions of years) by tectonic processes, which are responsible for the presence of the mountains in the first place, and in the shorter term (years to tens of millennia) by erosion processes. Erosion can be driven by cryospheric, hydrological, aeolian and biological factors, which are all ultimately governed by climate. Changing climate in a particular region of the world will lead to enhanced or decreased rates of erosion. Perhaps the only major non-climate erosion factor on relatively short time scales is earthquakes, but their consequences are generally limited to relatively small areas and hence earthquakes do not contribute as significantly as climate to the shaping of mountains and the natural systems which exist therein.

Much of the text which follows is based on research and data stemming essentially from mountain regions in Europe and North America, where proxy data are most abundant.

3.2 Proxy data: reconstructing the past

Mountain areas contain a variety of excellent natural archives that document former environmental changes with differing temporal and parameter resolution. These records involve both physical (e.g. glaciers, permafrost) and biological systems (e.g. trees) that respond to climate change and often occur in close juxtaposition, thereby providing complementary insights into former environmental changes. Since the time of Louis Agassiz, many key concepts about climate history have been developed from studies of the fluctuations of alpine glaciers (Penck and Bruckner, 1909; Matthes, 1939; Porter and Denton, 1967; Grove, 1987). Refining this history and supplementing it with new data, techniques and complementary sources such as tree rings, lacustrine records or historical sources allow a documentation of the magnitude and timing of past climate changes and benchmark natural climate variability at several time scales. Understanding and modelling this natural variability for periods ranging from decades to centuries is a critical element in our ability to detect any anthropogenic signal in the relatively short instrumental climate record. It defines the background variability and trends upon which any future climate changes will be superimposed. Studies of historical system responses, particularly over the past century, also allow us to develop potential analogues to predict the future response of these systems.

In many mountain environments (Stone, 1992), human activity has modified natural systems, such as tree lines in the Alps, making it difficult to discriminate between natural and anthropogenic causes of recent changes. In certain mountain regions such as the Canadian Rockies, for example, large areas were set aside within national parks prior to significant human impact, and observed changes can be attributed solely to natural forcing. In addition, the absence of significant human disturbance has allowed the preservation of physical evidence of change that has been destroyed elsewhere. However, this absence of human activity also limits access to the period of instrumental or documentary records: contrary to the situation in Europe, there are scant documentary observations for the Canadian Rockies prior to 1900, and few areas have been studied in detail. Greater reliance must therefore be placed on the development of natural archives to document recent climate history.

In order to obtain a perspective on climate variability prior to the late nineteenth century, when instrumental records started to become available, it is necessary to make use of proxy records of past climates and environments. Each type of record has its own attributes and limitations, and therefore it is advisable to look for a convergence of evidence before relying too much on one particular reconstruction.

Approaches to palaeoclimatic reconstruction in mountain areas are given here in a succinct manner, in order to provide a base for the subsequent sections on palaeoclimatic and palaeoenvironmental reconstructions.

3.2.1 Glacial geomorphology and glaciology

Numerous studies are available in the various mountain areas of the world on glacier variations and past glacier extent (Röthlisberger, 1986; Karlèn, 1988; Haeberli et al., 1990). The lower altitudinal limits of ice cover can be mapped and dated by ^{14}C (carbon-14 isotope) techniques for the last glaciation and by lichenometric methods for the past few centuries. For the more recent glacial advances or still stands, ice thickness, and therefore volume, may be estimated from trimlines on the valley sides. These changes in ice extent and inferred changes in snow-line altitude can be interpreted in terms of changes in ablation-season temperature and winter accumulation, although unique solutions cannot be derived. Some attempts have been made to model ice volume changes in terms of climatic forcings, such as the work of Allison and Kruss (1977) for Irian Jaya (New Guinea). More recent studies on glacier–climate modelling for the Alps have been carried out by Kuhn (1989), Oerlemans and Hoogendorn (1990) and others (see Oerlemans, 1989).

3.2.2 Ice cores

The extraction of climatic and other atmospheric signals from ice cores has considerable potential in high-altitude ice bodies above the zone of melting (Thompson et al., 1984). Falling snow traps air that is sealed off when ice is formed at about 50–100 m depth. The ice preserves records of atmospheric gas concentrations, aerosols, pollen and volcanic dust, as well as isotopic composition of the snow (related to crystallization temperature of the atmospheric water vapour) and net accumulation rate. In zones of high accumulation, annual layers can be distinguished for up to about 10,000 years, but in alpine glaciers the records span at best a few thousand years. Ice core studies in mountain areas include those on Quelccaya Icecap, Peru; Colle Gnifetti and Monte Rosa, Switzerland; and Dunde Icecap, western China (Wagenbach, 1989). Ice cores contain samples of past atmospheres, unaltered and in chronological order. They provide the only proven direct way to learn of pre-industrial atmospheric compositions as far back as 220,000 years ago. For example, ice-core data show that greenhouse gas concentrations have varied over the glacial–interglacial cycle in a way that can explain a significant fraction of the global temperature change (Barnola *et al.*, 1991). Ice cores provide the most faithful records of a wide range of particulate and chemical atmospheric impurities; similar records in lakes or oceans have been altered significantly by processes in the water prior to deposition. Ice cores show that the atmosphere contained more dust during the last ice age than today, which may have contributed to the observed cooling; records of methane–sulphate fluxes inferred from ice cores reveal variations in marine productivity that may have affected atmospheric CO_2 concentrations as well as the abundance of cloud-condensation nuclei; beryllium-10 fluxes reveal past solar activity; ammonia concentrations reveal past biomass burning; and nitrate concentrations may even provide a record of the Antarctic ozone hole (Oeschger and Langway, 1989). In addition, these cores provide the only records of local conditions over the 10 per cent of the land surface occupied by ice. Reconstruction of temperature and precipitation are often more accurate than those that can be obtained from other proxy methods in non-glaciated regions, allowing more rigorous model tests than would otherwise be possible.

3.2.3 Varved sediments

Attempts to link biological, geochemical, sedimentological or geophysical signatures of annually laminated (i.e. varved) sediments to climatic and hydrological parameters have been made by numerous researchers (e.g. Leonard, 1986; Teranes and McKenzie, 1995; Ohlendorf *et al.*, 1997). Pollen records extracted from lake sediments and peat bogs provide information on ecological changes at the genus, or occasionally the species, level for the past 10,000–30,000 years. However, pollen may be transported long distances, the composition of vegetation is not always fully or proportionately represented, and the ecological record must be converted to climatic information by some form of transfer function. For Pleistocene materials, an additional problem is caused when former plant assemblages lack a modern analogue. In mountain regions, local climatic characteristics complicate the problem of interpreting the local versus regional pollen rain (Markgraf, 1980). Thus, pollen embedded within the valley circulations may be isolated from that in the airflows above the ridge line. Further difficulties in the use of pollen data are the necessity to date the sediments by ^{14}C and the relatively coarse temporal resolution that is usually available, given the intervals between such dates, their standard errors, and variable sedimentation rates.

3.2.4 Dendrochronology

High-altitude forests are exceptional sites for dendrochronological studies, both in terms of climate reconstruction and for the assessment of the impacts of climatic change (Tessier *et al.*, 1997). The study of tree rings, namely their width and density, has several advantages. Dendrochronology allows temporal resolutions on annual and even seasonal scales over hundreds to a few thousands of years. Two distinct techniques provide climatic information, namely the measurement of annual ring width, and the analysis of wood density (Fritts, 1976; Schweingruber *et al.*, 1979). Ring width, when corrected for age trend, varies strongly in response to summer temperature towards polar and upper altitudinal tree lines, and to summer moisture near the lower altitudinal tree line in arid regions. In addition, the isotopic ratios (oxygen, deuterium and carbon) in tree rings are the result of the response of the tree to complex biological and ecophysiological processes, in which climate is probably the dominant actor. In order to obtain climate proxy data, it is necessary to disaggregate the climatic from the non-climatic factors responsible for the observed tree-ring characteristics. In all cases, the climatic factor is the only one which exhibits an annual periodicity and where the climatic anomalies are coherent over large regions. As a result, the interannual fluctuations of climate can be identified in the ring characteristics. Tree-ring studies can also be used to determine fire histories and volcanic eruptions, as well as stream-flow conditions using calibrated relationships between ring width and runoff amounts.

3.2.5 Historical records

Historical records have an absolute dating control, provide a high level of temporal resolution, from the daily to the yearly scales, and allow a disentangling of the effects of temperature and precipitation. Furthermore, as documentary data reflect the reaction of society to climatic events, they are highly sensitive to anomalies and natural hazards, and often include short-lived events such as hailstorms or severe frosts. The more extreme an event, the more often and the more completely it is described.

Documentary data do have a number of drawbacks associated with their interpretation, however. First, many of them are discontinuous and heterogeneous, and may be biased by the selective perception of the observer. Second, available compilations of documentary evidence tend to contain both reliable and unreliable data, which makes an analysis costly and time-consuming. In addition, the task of integrating material from numerous locations into a coherent body of evidence involves understanding the full range of languages and regional dialects which are spoken in Europe (Pfister, 1994). Finally, according to the nature of the data, the mathematical techniques of elaboration need to be relatively simple (Banzon et al., 1992), because much of the historical information other than that available in the form of continuous time series (e.g. vine harvest dates) is not suited for sophisticated statistical analyses.

Direct reports of weather-related conditions are contained in personal diaries and official records in European countries, China, Japan and elsewhere from the sixteenth or seventeenth centuries, as interest in natural science developed, or even earlier in association with documentation of agricultural production for taxation purposes. In the Alpine countries, Britain and Scandinavia, some of these sources provide information on the climatic deterioration in mountain and upland areas during the Little Ice Age compared with the Medieval Warm Period (Parry, 1978; Pfister, 1985a,b; Grove, 1987; Zumbühl, 1988).

3.2.6 The instrumental record

Mountain observations can provide an indication of changes in the free atmosphere for a much longer period than sounding data, and measurements on the Zugspitze (Bavarian Alps, Germany), on the Sonnblick (Austria) and in the Swiss Alps provide such long-term series. However, at many of these stations there are problems of data homogeneity due to changes of instrumentation or observing practice. Records from mountain observatories enable a comparison of the relative amplitudes of trends in the mountains and on adjacent lowlands.

Systematic observations at mountain observatories did not begin until the late nineteenth century (Barry, 1992a). Diaz and Bradley (1997) identify some 22 stations in Europe, 8 in Asia, and 37 in the USA above 1500 m with long-term records, i.e. longer than 30 years. However, not all of these are mountain stations, because some are on high plateaux and others are in mountain valleys. Data are also available from numerous second-order and auxiliary stations. These include, for example, simple weather-recording stations which were set up in the vicinity of hydro-power construction sites.

Intuitive reasoning suggests that the further back one delves into the past, the more difficult it becomes to reconstruct palaeoenvironmental conditions, because either the information is missing or it has been transformed by more recent events. For example, in the case of climate reconstruction using varved sediments, these type of proxy records provide valuable information only since the end of the last glaciation in regions which were covered by icecaps at the time. This is because glacier and subsequent fluvial erosion have removed or significantly perturbed the stratified sediments in lakes which existed prior to the last major ice age.

3.3 Environmental change in the distant past

Figure 3.1 illustrates schematically the major climatic events which have occurred over the past 5 million years. Although the diagram shows climatic conditions which hold for the entire globe, mountain environments were inevitably under the general influence of these climates. As a result, the distribution of ecosystems, snow and ice, and water would have been substantially different according to

the degree of warmth or cold with respect to current conditions. It is interesting to note that over the period illustrated in this figure, the Earth entered into a period of relative rapid fluctuations from ice ages to interglacial phases with a frequency which did not exist prior to about 3 million years BP.

The *Pliocene Optimum* is a warm period which occurred between 4.3 and 3.3 million years BP and terminated with the beginning of the more frequent glacial/interglacial periods. Much of the information on this Optimum comes from deep ocean sediments and is essentially confined to the northern hemisphere. According to the available proxy data, mid-latitude summer temperatures were 3–4°C higher than today, and atmospheric CO_2 concentrations at about 600 ppmv (compared to pre-industrial values of about 280 ppmv). Indices from the Saharan zone and other arid regions such as Central Asia indicate that precipitation was generally more abundant in North Africa, where summer temperatures

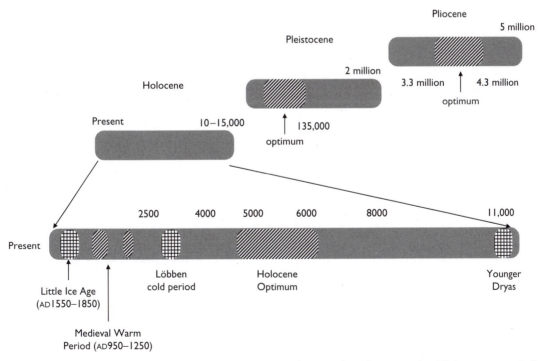

FIGURE 3.1 Schematic illustration of climatic events during the Pliocene, the Pleistocene and the Holocene. Dates are in years BP

were lower than today. Vegetation and animal life were considerably different, with species responsive to warm conditions far more abundant at higher mountain elevations and at higher latitudes than today.

Difficulties arise when attempting to provide a clearer picture of this period. Inaccuracies in dating methods for such a distant past imply a margin of error of about 100,000 years. Geographical conditions were also significantly different from today. For example, the land–ocean distribution was different from that today; there was much less ice at the surface (especially in Greenland), so that sea level was substantially higher than at present. The Tibetan Plateau was about 1000 m lower than its present average altitude, which certainly had implications for climatic processes that are linked to the presence of the Himalayas and Tibet (for example, the Indian monsoon).

The Pliocene Optimum is of interest to climate specialists working in the field of human-induced climatic change (*see* Chapters 5 and 6) because projected climatic change in the twenty-first century is likely to lead to similar levels of warming and CO_2 concentrations. Analogies with the past are difficult, however, because of the lack of available data, and the fact that direct human environmental stress did not exist in the Pliocene, and because the rates of change expected in the twenty-first century are one to two orders of magnitude greater than natural climatic fluctuations.

The Eemian period is a climatic optimum which occurred during interglacial phases of the Pleistocene, centred roughly around 135,000 years BP. During this warm phase, atmospheric CO_2 levels of 300 ppmv were close to today's. The eccentricity of the Earth's orbit was greater than today, leading to greater amounts of solar radiation affecting the northern hemisphere in summer, yielding global average temperatures 2°C warmer than today. However, in boreal latitudes, temperatures were 4–8°C warmer in Siberia, northern Canada and Greenland, 1–3°C warmer in the 50–60° latitude belt in the former Soviet Union, while further south, temperature conditions were similar to today's.

Overall, precipitation in the northern hemisphere was more abundant than now. The natural environment of many mountain regions must have been similar to today's, because climatic conditions were not vastly different from current conditions.

Uncertainties linked to this period are related to lack of available proxy data in many parts of the world, in particular in arid regions, where an enhanced precipitation signal is likely.

3.4 Mountain environments during the last major glaciation

For the last glacial cycle, global climates have been reconstructed by several groups using a variety of palaeoclimatic data extracted from cores of ocean sediments, lake and bog sediments, ice layers, tree rings, etc. (CLIMAP Project Members, 1976, 1981; COHMAP Members, 1988). Ocean temperatures estimated for the last glacial maximum (LMG), about 18,000 BP based on planktonic foraminifera, show either modest cooling (~2°C) or even warming of ocean surface waters in most low- and subtropical-latitude areas. In contrast, glacial geomorphologic evidence of former glacier extent and downward displacement of vegetation zones in tropical mountain areas of the order of almost 1 km imply substantially greater cooling of 5–6°C. This pattern has, for example, been found for New Guinea (Bowler *et al.*, 1976) and Hawaii (Porter, 1979).

Webster and Streten (1978) and Rind and Peteet (1985) discuss the evidence and possible explanations, if the interpretations of the palaeoclimatic records are assumed to be reliable. They suggest several physical causes: the atmospheric temperature lapse rate may have been steeper than now, or more frequent incursions of upper-level cold troughs from middle latitudes may have increased snowfall and lowered temperatures. Simulations of snow-line lowering using the CLIMAP-reconstructed sea surface temperature data and Goddard Institute for Space Studies climate model (8° latitude × 10° longitude resolution) indicated lowerings only half (or less) of those observed, implying that the CLIMAP data may have a warm bias (Rind and Peteet, 1985). Subsequently, Rind (1988) has shown that model results must be used with great caution. By using a finer grid resolution (4° × 5°) than in the previous study, he demonstrates that more atmospheric cooling and snow-line lowering is simulated for the LGM in the Hawaiian region, although little change occurred in the other tropical mountain areas (Colombia, New Guinea and East Africa).

3.5 Mountain environments during the Holocene

3.5.1 Changes in physical systems

The Holocene spans the time-frame from the termination of the last major glaciation period, at about 15,000 BP, to today. Because of the greater availability of proxy data and the enhanced confidence one has in such information, it is possible to identify a number of periods where conditions were either colder or warmer than the twentieth century.

In the highest mountains, icecaps and glaciers are dominant features, and in a few areas, ice cores have been recovered from mountain icecaps, providing a set of unique high-resolution records extending back for approximately 100 to over 10,000 years. High-elevation ice cores provide invaluable palaeoenvironmental information to supplement and expand upon that obtained from polar regions (Greenland and Antarctica, essentially). To date, four high-altitude sites have yielded ice cores to bedrock: Quelccaya and Huascarán in Peru, and Dunde and Guliya Icecaps in western China. Where records extend back to the last glacial period, which is indeed the case in the Dunde and Huascarán ice cores, glacial-stage ice is thin and close to the base, making a detailed interpretation very difficult (Thompson *et al.*, 1988a, 1989, 1990, 1995). Nevertheless, even short sections of deep ice cores can yield important information. For example, in ice cores from Huascarán, Peru, located at 6048 m above sea level, the lowest few metres

contain ice from the last glacial maximum, with the isotopic ratio $\delta^{18}O$ approximately 8 per cent lower than Holocene levels, and a much higher dust content (Thompson *et al.*, 1995). The lower $\delta^{18}O$ suggests that tropical temperatures were significantly reduced in the LGM by about 8–12°C), which supports arguments that changes in tropical sea surface temperatures (SSTs) were much greater than those indicated by the reconstructions of CLIMAP (1981) which have guided thinking on this matter for many years.

Because of the high accumulation rates on mountain icecaps, high-elevation ice cores can provide a high resolution record of the recent past, with considerable detail on how climate has varied over the past 1000–2000 years in particular (Thompson, 1991, 1992). The Quelccaya ice cores have been studied in most detail over this interval (Thompson *et al.*, 1985, 1986; Thompson and Mosley-Thompson, 1987). Certain cores which extend back almost 1500 years reveal a fairly consistent seasonal cycle of microparticles, conductivity and $\delta^{18}O$ which has been used to identify and date each annual layer. Dust levels increase in the dry season (June to September) when $\delta^{18}O$ values and conductivity levels are highest, providing a strong annual signal.

$\delta^{18}O$ over the past 1000 years shows distinct variations in the Quelccaya core, with lowest values from AD 1530 to 1900. Accumulation was well above average for part of this time (1530–1700) but then fell to low levels. The overall period corresponds to the Little Ice Age observed in many other parts of the world. Accumulation was higher prior to this interval, especially from about AD 600 to 1000. Archaeological evidence shows that there was an expansion of highland cultural groups at that time. By contrast, during the subsequent dry episode in the mountains (AD 1040–1490), highland groups declined, while cultural groups in coastal Peru and Ecuador expanded (Thompson *et al.*, 1988b). This may reflect longer-term evidence for conditions which are common in El Niño years, when coastal areas are wet at the same time as the highlands of southern Peru are dry. Indeed, the Quelccaya record shows that El Niños are generally

Box 3.1   Evidence for glacier retreat in the Alps during the Holocene

The range of pre-industrial and prehistoric variability is well recorded by the behaviour of glaciers. The reconstruction of past glacier length changes is based on direct measurements, old paintings, written sources, moraines, pollen analysis, tree-ring investigation, etc. These records indicate that the volume and surface area of Alpine glaciers was less than today, and that the rates of change observed during the twentieth century were probably not uncommon during the Holocene (Zumbühl and Holzhauser, 1988). During Roman times, for example, mule tracks led from Italy to the Rhone Valley in Switzerland over ridge crests above Zermatt, which are today sufficiently glaciated to allow summer skiing in the region. On the other hand, Alpine glacier extent has varied over the past millennia within a range defined by the extremes of the Little Ice Age maximum extent and today's reduced stage (Gamper and Suter, 1982). This implies that there is an evolution of glaciers towards or even beyond the 'warm' limit of natural Holocene variability. These allow a disentangling of the strong and rapidly increasing human impact on the climate against the background of externally or internally forced fluctuations.

Most recently, evidence has also emerged from sites other than the frontal zone of glaciers, i.e. from the top of glacier accumulation areas. Even at low altitudes, wind-exposed ice crests are not temperate but are slightly below the freezing point at their bases, and as a result adhere to the underlying bedrock (Haeberli and Funk, 1991). Such conditions at the glacier base – i.e. reduced heat flow through winter snow, no melt-water, no basal sliding, low to zero basal shear stress at interface between permanent snow and ice – explain the perfect conservation of the 'Oetztal Ice Man', whose body had been buried beneath the ice in a small topographic bedrock depression on a saddle ridge at Hauslabjoch (Austrian Alps; 3200 m above sea level) more than 5000 years ago. The intact body thereafter remained in place until it melted free in 1991 (VAW, 1993). At an even lower altitude (2700 m above sea level) but at a comparable site (Lötschenpass, Swiss Alps), three well-preserved wooden bows and a number of other archaeological artefacts were discovered in the first half of the twentieth century. Recent carbon-14 dating of the three bows gave dendro-chronologically corrected ages of around 4,000 years (Bellwald, 1992). These archaeological findings confirm that glacier length variation and overall glacier mass change indeed occur simultaneously if considered at the century time scale. Most essentially, however, the recent evidence of prehistoric settlements from melting ice in saddle-type orographic configurations imply that the extent of glaciers and permafrost in the Alps may be more reduced today than since the Holocene Optimum around 5000 years BP.

associated with low-accumulation years, though there is no unique set of conditions observed in the ice core which permits unequivocal identification of an ENSO event (Thompson *et al.*, 1984). Nevertheless, by incorporating ice-core data with other types of proxy record it may be possible to constrain long-term reconstructions of ENSO events (Baumgartner *et al.*, 1989; Michaelson and Thompson, 1992).

High-altitude ice cores reflect significant increases in temperature over the past few decades, resulting in glaciers and icecaps disappearing altogether in some places (e.g. Schubert, 1992). This is quite different from polar regions, where temperatures have declined in many regions during the same period. At Quelccaya, temperatures in the past 20 years have increased to the point that by the early 1990s melting had reached the Summit core site (5670 m), obscuring the detailed $\delta^{18}O$ profile that was clearly visible in cores recovered in 1976 and 1983 (Thompson *et al.*, 1993). In the entire 1500-year record from Quelccaya, there is no comparable evidence for such melting at the Summit site. Similarly, at Huascarán, in northern Peru, $\delta^{18}O$ values increased markedly, from a Little Ice Age minimum in the seventeenth and eighteenth centuries, reaching a level for the past century which was higher than at any time in the past 3000 years. Ice cores from Dunde Icecap, China, also show evidence of recent warming; $\delta^{18}O$ values are higher in the past 50 years than in any other 50-year period over the past 12,000 years (though the resolution of short-term changes in amplitude decreases with time). These records, plus evidence from other short ice cores from high altitudes (Hastenrath and Kruss, 1992), point to a dramatic climatic change in recent decades, prompting concern over the possible loss of these unique archives of palaeoenvironmental history (Thompson *et al.*, 1993).

Ice marginal positions are recorded by moraines, and trimlines on valley walls. In ideal situations, it may be possible to identify a series of overlapping or nested moraines representing former glacier positions. However, more commonly the most recent advance of ice (the exact timing of which may have varied) has obliterated evidence of earlier advances because it was the most extensive for several millennia (and in some glaciers, the most extensive since the last ice age). These Little Ice Age moraine systems are the latest in a series of glacier advances which began in the late Holocene, and which are collectively referred to as *neoglaciations* (Grove, 1987; Matthews, 1991). Dating such advances is problematical, relying principally on radiocarbon dating of organic material buried by the advancing moraine, or by lichens growing on the moraine itself, once it has stabilized. Clearly, such evidence can only be episodic and does not provide the kind of high-resolution, continuous data that is favoured in palaeoclimatic analysis. However, closely related records may be obtained from

glacier-fed lakes, which may register the growth of ice and the deterioration of mountain climates by a reduction in organic matter, and an increase in silt input to the lake bottom sediments (Karlèn, 1976, 1988; Nesje *et al.*, 1991). In ideal circumstances, such records may be annually laminated, providing very high-resolution insight into past climatic conditions (e.g. Leonard, 1986). Pollen and other microfossils in lake sediments, or in high altitude bogs, can be interpreted in terms of former tree-line movements and hence provide a framework for other proxy records in the mountains. On the basis of a composite view of such data, Karlèn (1993) argues that glaciers (in the more continental parts of Scandinavia) advanced to positions comparable to those of the Little Ice Age around 3000, 2400, 2000 and 1200 years BP. Many of these glaciers had completely disappeared in the early to mid-Holocene, re-forming only within the past 3000 years (Matthews, 1993; Nesje *et al.*, 1994).

Beyond the realm of snow and ice, and alpine tundra, the tree line defines an important climate-related ecotone. Although the tree line itself varies in structure and composition from one mountain region to another, and is subject to many potentially limiting ecological constraints (Tranquillini, 1993), climate is the dominant control, at least away from the oceanic margin. Consequently, evidence of past changes in tree-line position is generally interpreted in terms of variations in summer temperature. Radiocarbon-dated macrofossils (tree stumps or wood fragments) from above the modern tree line can thus provide testimony of warmer conditions in the past. This is well illustrated in the northern Urals, where now-dead trees beyond the modern tree line have been dated using dendrochronology techniques to obtain information on the timing of past tree growth at high elevations. This reveals that most of the trees were growing in the tenth to twelfth centuries AD; no trees were found to date from the late eighteenth and nineteenth centuries, indicating that the tree line had retreated at that time. This evidence is strongly supported by tree-ring studies in nearby forests, where maximum ring widths were found at the time the forest advanced, and minimum ring widths were characteristic of the eighteenth and nineteenth centuries.

Sub-fossil wood from above the present tree limit has been found over wide areas of Scandinavia, and the mountains of the western United States (Kullman, 1989, 1993; Kvamme, 1993; Rochefort et al., 1994). In both areas, there is strong evidence that the upper tree limit was well above modern levels, especially before about 5000 BP. In parts of the western United States and western Canada, trees were growing as much as 150 m above modern limits in the period from 8000 to 6000 BP (Rochefort et al., 1994), and similarly, in Scandinavia, trees were up to 300 m above modern limits in the early Holocene, suggesting that summer temperatures were 1.5–2°C above modern levels (Karlèn, 1993; Kvamme, 1993). In both areas, it appears that climate deteriorated after 5000 BP, leading to a reduction in tree limits. This corresponds to both pollen records and the evidence from glacier moraines that temperatures became lower, especially after about 3500 BP, marking the onset of late Holocene neoglaciation. Minor oscillations of tree-line have taken place since then, culminating in the coldest episodes in the sixteenth to the mid-nineteenth centuries. At that time, temperatures in the mountains of southern Sweden were around 1°C colder than in the mid-twentieth century (Kullman, 1989).

The overall picture from diverse palaeoclimatic records in mountain areas is thus of early Holocene warmth, reaching an optimum around 6000 BP, followed by a cooler late Holocene. The period after 5000 BP was punctuated by a few warm periods but there were also several especially cold episodes when glaciers advanced and tree line declined. In terms of the current debate over anthropogenic versus natural climate forcing, it is important to note that the instrumental record which is now used to characterize 'global warming' began at what was arguably the coldest period of the Holocene, in the mid-nineteenth century. Clearly, pronounced climatic variations have been registered by proxy records in mountain areas long before any significant anthro-

pogenic effects on global greenhouse gas concentrations, yet the causes of such variations remain obscure. Wigley and Kelly (1990) and Magny (1995) point out the correlation between Holocene glacier fluctuations and ^{14}C variations, a proxy which they consider to be indicative of solar irradiance variations. However, the ^{14}C anomaly record is influenced by several factors, including changes in deep water circulation of the ocean; it may be that the apparent links between ^{14}C and glacier fluctuations reflect subtle changes in the thermohaline circulation of the ocean, which will clearly influence the atmospheric circulation on a global scale. There are others who believe that the record of mountain glacier fluctuations (at least for the past 1000 years) is closely linked to the level of volcanic aerosols in the atmosphere, and by inference, similar events earlier in the Holocene may also be indicative of higher volcanic dust loading of the atmosphere (Nesje and Johannessen, 1992). Whichever of these factors has been important, there is little doubt that over a very long time-frame, changes in the amount of solar radiation intercepted at the Earth's surface, related to changes in the Earth's orbital parameters, have been the dominant factor controlling mountain climates over the past 20,000 years, and it is probable that the observed record of higher tree line in the early Holocene is directly attributable to increased solar radiation and warmer summer temperatures at that time. The fairly rapid shift to lower tree line, and the onset of neoglaciation soon after 5000 BP in many areas, seems to have been too abrupt to be simply due to declining radiation receipts, and some of the other factors mentioned may have contributed, individually or collectively, to the cooling that occurred at that time. Later cool episodes, culminating in the series of cold spells of the Little Ice Age (sixteenth to nineteenth centuries AD), may be related to reduced solar irradiance and/or enhanced volcanic activity, though the possibility of a reduction in North Atlantic thermohaline circulation that principally affected Europe may also be a contributing factor (Rind and Overpeck, 1993; Keigwin, 1996).

3.5.2 Changes in biospheric systems

Climatic variability is an important exogenous factor affecting regeneration dynamics and successional processes in forest communities (Davis, 1986; Prentice, 1986). Both modelling and palaeoecological approaches have proven useful in elucidating the effects of climate change on vegetation dynamics. Current capability to predict climate-induced vegetation change is limited, however, by spatial and temporal scale problems (Neilson, 1986), and inadequate understanding of the effects of climate variation on disturbance regimes (Graham *et al.*, 1990; Overpeck *et al.*, 1990). Separating climate-induced from human-caused vegetation changes represents a major challenge to properly assess the influence of climatic variations on vegetation dynamics.

The interrelated systems of climate, vegetation and physical landscape are highly dynamic on time scales ranging from hours to millions of years. The control of vegetation patterns by climate is most obvious at coarse spatial and temporal scales (Webb, 1987), but at time scales of decades to several centuries, the effects of climatic variation on vegetation dynamics have been less obvious. This is often due to the difficulty of separating disturbance-induced successional changes from climate-induced changes (Veblen and Stewart, 1982; Ogden, 1985; Brubaker, 1986; Davis, 1986; Veblen and Lorenz, 1987, 1988; Veblen and Markgraf, 1988; Prentice, 1992; Archer *et al.*, 1995). However, it is precisely at this time scale of years to decades that the understanding of the effects of climatic variation on vegetation change is so vital (Melillo *et al.*, 1996; Overpeck *et al.*, 1990; Solomon, 1986).

Most studies of the effects of climatic variation on forest dynamics have focused on tree regeneration (Brubaker, 1986), and few have incorporated the effects of climatically influenced disturbance regimes (e.g. Bradshaw and Zackrisson, 1990; Clark, 1988, 1990; Grimm, 1983; Johnson and Larsen, 1991). Furthermore, the effects of climate on tree mortality have rarely been considered (Betancourt *et al.*, 1993; Elliot and Swank, 1994; Szeicz and MacDonald, 1995). As the death of mature trees is important in releasing resources potentially available for the new establishment of trees of the same or of different species (Franklin *et al.* 1987), determining the influences of climatic variation on tree mortality is essential for understanding the effects of climatic variation on forest dynamics.

Trees respond to decadal-scale climatic variation by systemic changes in carbon balance and nutrient status which, in turn, lead either to changes in absolute growth rates or to altered patterns of carbon allocation. Two broad classes of response have occurred in coniferous species in response to the temperature fluctuations of the last millennium. First, at several alpine and Arctic tree lines, range limits have shifted owing to altered reproductive and establishment rates. Second, in areas where climatic limitations to growth have changed, established trees have undergone phenotypic adjustments to the altered climate.

While climatic variables may subtly alter rates of reproduction and establishment in temperate coniferous ecosystems, the role of climate in governing population processes is clearly seen at alpine and Arctic ecotones. Tree establishment has increased in subalpine and tree-line stands, and young trees have established at elevations or latitudes beyond the current tree line in a variety of settings worldwide. However, ecotonal movement is not universally observed in all regions that have experienced a recent upward trend in temperatures.

Sustained changes in growth and productivity have been observed during the twentieth century in trees growing near latitudinal and altitudinal tree line in western North America (Garfinkel and Brubaker, 1980; LaMarche *et al.*, 1984; Graumlich and Brubaker, 1986; Graumlich, 1991; Graybill and Idso, 1993; Luckman, 1994). Growing-season temperature is generally the most important climatic variable governing growth at latitudinal and altitudinal tree line. However, analyses of year-to-year variability indicate that other climatic factors, such as soil moisture and depth of snow pack, interact with temperature in controlling growth at tree line sites in the Cascade Mountains and Sierra Nevada

Box 3.2 Dendrochronological studies of fluctuations in tree growth in Patagonia, Argentina

A study by Villalba and Veblen (1998) has explored the role that climatic fluctuations at annual to decadal scales have on *Austrocedrus chilensis* establishment and mortality. These trees are located at the forest–steppe border in northern Patagonia, Argentina, from 37 to 43°S. Recent episodes of tree establishment and mortality have been reconstructed on the basis of age frequency distributions of living trees as well as dendrochronological dating of snags.

Austrocedrus establishment at the forest–steppe ecotone appears to be episodic in relation to climatically distinct episodes. Wet and cool summers prevailing for a decade or longer facilitate tree establishment. For the past 150 years, tree establishment at the forest–steppe ecotone in this region of Argentina has involved interactions between climate and human disturbances, principally those linked to fire and grazing. Episodic mortality is generally associated with extremely dry and warm climatic events during a single summer or during two consecutive summers. Most episodes of tree establishment and mortality are concurrent with years of above- and below-average tree growth, respectively. However, regional episodes of *Austrocedrus* establishment are related to decadal-scale climatic events, whereas mortality episodes are mainly controlled by extreme, annual-scale climatic fluctuations.

A period of sustained radial tree growth above the long-term mean occurred from 1868 to 1892. Two subperiods, 1868–74 and 1887–92, have tree-ring indexes similar in magnitude to those recorded during the most recent period of massive establishment in the 1970s (Figure 3.2). The difference in the mean amplitude of the first principal component for the interval 1868–92 versus 1964–77 is not great enough to reject the possibility that the two intervals are subsamples

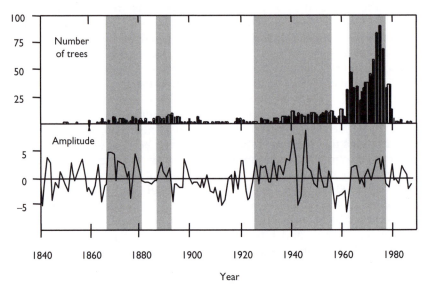

FIGURE 3.2 (a) The age–frequency distribution from 1312 *Austrocedrus* trees along the forest–steppe ecotone in northern Patagonia, Argentina, and (b) the radial tree growth summarized as the amplitude of the first principal components from 15 tree-ring chronologies covering the interval 1800–1989

from the same population. The increase in abundance of trees starting in the 1860s and lasting until about 1900 may reflect favourable climatic conditions.

The interval 1899–1917 is the most extended period of below-average tree growth over the past 300 years (Villalba, 1995; Villalba and Veblen, 1997a,b). Concurrent with this period of below-average growth, the number of surviving trees declined (Figure 3.2). Tree-growth indexes gradually increased during the 1930s, peaked in 1941 and suddenly decreased to extremely low values in 1943 and 1944. The growth indexes from 1933 to 1941 represent the most conspicuous peak in tree growth recorded in the twentieth century. Although a slight increase in the number of surviving trees occurs after 1936, the magnitude of this change is not great relative to the increase in radial growth. However, the extremely hot and dry summers of 1942 and 1943 correspond to a regionally extensive episode of mortality of mature trees, during which an increase in seedling mortality seems to have occurred. Thus, any favourable response of tree populations to the moister conditions of the 1930s may be obscured by increased mortality of young trees in the early 1940s.

Tree-ring growth decreased to its lowest values during the dry, warm period from 1957 to 1962. This period is also a time of relatively scarce surviving trees, at least in comparison with the following period of increased radial growth, and abundant surviving trees that began around 1963. The increase in surviving trees in the 1963–79 period parallels an increase in tree radial growth. Radial tree growth was relatively

low in 1978–79 but has increased since 1980 and remained above the mean between 1984 and 1987. The relationship between tree growth and climate appears to have changed during the 1980s, so that despite the warm and dry conditions indicated by the instrumental record, tree growth did not decline to any large extent (Villalba, 1995). However, this period of warm and dry conditions is clearly a time of scarce tree establishment.

Although the age structure patterns of *Austrocedrus* strongly coincide with climatic variation for the period since the early 1960s, for earlier time periods the influence of climatic variation is less obvious. For example, the tree-ring record indicates that the 1860s to 1890s and 1930s to early 1950s were characterized by spring and summer moisture availabilities at least as great as those of the 1963–79 period of abundant tree survival. However, the regional tree age structure indicates only modest increases in the abundances of trees surviving from these earlier periods of favourable climate. Owing to the cumulative effects of mortality on older cohorts, it is not surprising that past periods of favourable climate become increasingly difficult to detect as the cohort ages. For example, as the 1962–79 cohort ages and mortality reduces the abundance of surviving trees, the recognition of this period as one of abundant tree establishment will become more difficult. Over the longer term, variations in disturbance regimes, some related to climatic variations and others not (e.g. fire versus grazing), have also undoubtedly influenced the regional pattern of *Austrocedrus* age structures (Veblen *et al.*, 1992).

(Graumlich and Brubaker, 1986; Graumlich, 1991).

The two climatic responses documented above – increased regeneration rates and increased growth rates in tree line stands – appear unique when considered in the context of twentieth-century observations. However,

in the relatively rare circumstances in which particularly long-lived conifers provide growth or regeneration records that extend over the past millennium, the responses to recent climatic trends are not unique and have occurred in the past when temperatures were equivalent to late twentieth-century values. Similarly, evidence for tree establishment above the current tree line exists at several localities throughout the world, indicating that the climatic variables critical for regeneration and survival have fluctuated significantly on the time scales of hundreds of years to millennia. Therefore, while the warm temperatures of the twentieth century may be anomalous with respect to the life span of most coniferous species, they may not be anomalous in the context of the recent history of a given forest stand.

The ecological situation at the timber line is particularly complex in areas where anthropogenic disturbances interfere with natural factors, as is the case in the European Alps and many other high mountain ranges of Eurasia that have been settled since prehistoric times. In the Alps, for example, the timber line was lowered by 150–400 m with respect to its uppermost position during the post-glacial thermal optimum. This was in response to alpine pasturing, local mining and salt works during the Middle Ages (Holtmeier, 1986). As a consequence of the extensive deforestation of high-elevation areas, avalanches, landslides, soil erosion and torrential washes became more frequent and a permanent threat to populations and settlements in the mountain valleys.

At present, the climatic limit of tree growth is located above the actual forest limit, as is clearly evidenced by the invasion of trees into abandoned or rarely used alpine pastures. This invasion has certainly been enhanced by the warming in the second half of the twentieth century. The anthropogenic timber line has become as pronounced an ecological boundary as the natural timber line had been before (Holtmeier, 1994).

3.6 Climatic change in the twentieth century

3.6.1 Analyses of climatic characteristics in the European Alps

The climate of the Alpine region is characterized by a high degree of complexity due to the interactions between the mountains and the general circulation of the atmosphere, which result in features such as gravity wave breaking, blocking highs and föhn winds. A further cause of complexity is inherent to the Alps as a result of the competing influences of a number of different climatological regimes in the region: Mediterranean, continental, Atlantic and polar. Traditionally, the Alps are perhaps the best-endowed mountain system in terms of climatological and environmental data, and it is here that many of the most relevant studies of climate and climate change in mountain regions have been undertaken.

Figure 3.3 shows the changes in yearly mean surface temperature anomalies during the twentieth century from the 1951–80 climatological mean, averaged for eight sites in the Swiss Alps. These observational sites range in altitude from 569 m above sea level (Zürich) to close to 2500 m (Säntis). The global data of Jones and Wigley (1990) have been superimposed here to illustrate the fact that the interannual variability in the Alps is more marked than that on a global or hemispheric scale; the warming experienced since the early 1980s, while synchronous with the global warming, is of far greater amplitude and reaches up to 1°C for this ensemble average and up to 2°C for individual sites, such as the Säntis. This represents roughly a five-fold amplification of the global climate signal in the Alps (Diaz and Bradley, 1997). Similar studies for the Austrian Alps (e.g. Auer et al., 1990) and the Bavarian Alps, based on climatological records from the Sonnblick Observatory (Austria) and the Zugspitze (Germany), lead to broadly similar conclusions (Weber et al., 1994).

Beniston et al. (1994) have undertaken an exhaustive study of climate trends this century in the Swiss Alps. Climate change in the region during this period was characterized by increases in minimum temperatures of about 2°C, a more modest increase in maximum temperatures (in some instances a decrease of maxima in the latter part of the record), little trend in the precipitation data, and a general decrease of sunshine duration through to the mid-1980s. Warming was most intense in the

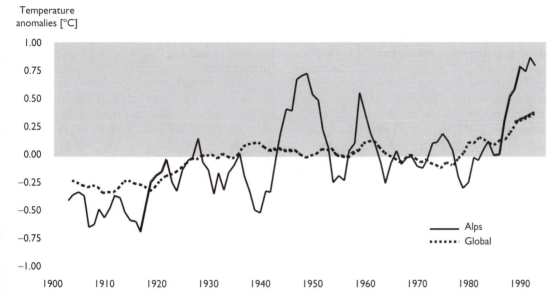

FIGURE 3.3 Temperature anomalies from the 1951–80 climatological mean in the Swiss Alps compared to the global anomalies

1940s, followed by the 1980s; the cooling which intervened from the 1950s to the late 1970s was not sufficient to offset the warming in the middle of the century. The asymmetry observed between minimum and maximum temperature trends in the Alps is consistent with similar observations at both the subcontinental scales (e.g. Katz and Brown, 1992, for trends in the Colorado Rocky Mountains and neighbouring Great Plains locations) and the global scale. Karl *et al.* (1993) have shown that over much of the continental land masses, minimum temperatures have risen at a rate three times faster than the maxima since the 1950s, the respective anomalies being of the order of 0.84°C and 0.28°C during this period.

Pressure statistics have been compiled as a means of providing a link between the regional-scale climatological variables and the synoptic, supra-regional scale. These statistics have shown that pressure exhibits a number of decadal-scale fluctuations, with the appearance of unusual behaviour in the 1980s; in that particular decade, pressure reached annual average values far higher than at any other time during the twentieth century. The frequency of occurrence of high-pressure

episodes exceeding a sea-level reduced threshold of 1030 hPa between 1983 and 1992 accounts for one-quarter of all such episodes over the past century. The pressure field is well correlated with the North Atlantic Oscillation (NAO) Index – a measure of the strength of flow over the North Atlantic, based on the pressure difference between the Azores and Iceland – for distinct periods of the Swiss climate record (1931–50 and 1971–90), and shows almost no correlation with the NAO Index for the other decades of the century. This is indicative of a transition from one climatic regime to another, dominated by zonal flow when the correlation with the NAO Index is high (Figure 3.4). In the 1980s, when zonal flow over the North Atlantic was particularly strong, episodes of persistent, anomalously high pressures were observed over the Alps and in southern Europe, particularly during the winter season (Hurrell and van Loon, 1997). This resulted in large positive departures in temperature anomalies and a significant lack of precipitation, particularly in the form of snow, in the Alps. Indeed, much of the strong warming observed since the 1980s and illustrated in Figure 3.3 can be

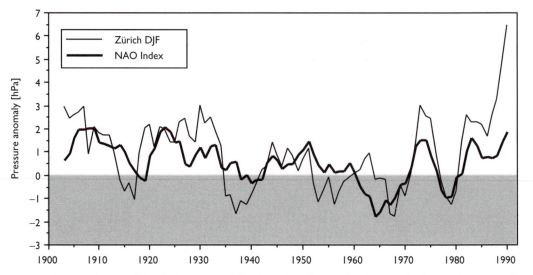

FIGURE 3.4 Comparisons of the behaviour of the North Atlantic Oscillation Index and mean winter (DJF: December, January and February) pressure in Zürich, Switzerland

explained to a large degree by these persistent high-pressure anomalies. During this same period in Europe, the Iberian Peninsula was under the influence of extended periods of drought and very mild winters, while northern Europe and Scandinavia were experiencing above-average precipitation (Hurrell, 1995) and, in certain cases, spectacular advances by Norwegian glaciers.

Beniston and Rebetez (1996) have shown that temperature anomalies exhibit not only a time-dependency on observed warming, but also an altitudinal dependency. Taking the 15-year period from 1979 to 1994, they have shown that minimum temperature anomalies exhibit a significant linear relationship with altitude, except at low elevations which are subject to wintertime fog or stratus conditions; the stratus or fog tends to decouple the underlying stations from processes occurring at higher altitudes. The same authors have also shown that there is a switch in the gradient of the temperature anomaly with height from cold to warm winters. For warm winters, which represent seven years in the 15-year sample, the higher the elevation, the stronger the positive anomaly; the reverse is true for cold winters (accounting for five winters during this period). This is illustrated in

Figure 3.5 for 39 Alpine stations. The figure provides a measure of the damping of the climate signal as an inverse function of height. The spread of data is the result of site conditions: topography plays a key role in the distribution of temperature means and anomalies, in particular for stations located on valley floors. In such sites, temperature inversions frequently occur and lead to a decoupling of temperatures at lower levels from those above the inversion, which are closer to free atmosphere conditions. Recent work by Giorgi *et al.* (1997), based on high-resolution simulations with a limited-area model, has provided results which are broadly consistent with the statistical analysis of Swiss data. In many general circulation models and regional climate model simulations, the altitudinal dependency of temperature anomalies can be observed (Vinnikov *et al.*, 1996; Diaz and Bradley, 1997).

One of the numerous problems associated with future climate change induced by anthropogenic emissions of greenhouse gases is the behaviour of extreme temperatures and precipitation and how these may respond to shifts in the means (Giorgi and Mearns, 1991; Katz and Brown, 1992; Beniston, 1994; IPCC, 1996). Extreme climatological events tend to have a

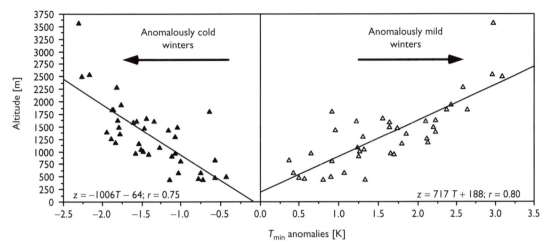

FIGURE 3.5 Illustration of the height dependency of temperature anomalies for cold and warm winters in the Swiss Alps

greater impact on natural and socio-economic systems than changes in mean climate, which is the commonly used concept for discussing climatic change, on a wide range of environmental and socio-economic systems. On the basis of the high spatial and temporal density of the Swiss climatological observing network, it has been possible to evaluate how shifts in climate during the twentieth century influenced the extremes of temperature (Beniston, 1997) at different locations in the Alpine region. Their study has shown that temperature extremes were observed to shift by a factor of over 1.5 for a unit change in means during the century; there is additionally an asymmetry between the shifts in the lower (minimum) extreme and the upper (maximum) extreme, leading to a changed frequency distribution profile for temperatures. This is illustrated in Figure 3.6 for the climatological station of Davos, located at 1590 m above sea level in the eastern part of the Swiss Alps.

Precipitation in the warmer part of the climatological record (i.e. from 1980 to the present) has been shown to exhibit a bimodal shift towards drier conditions, on the one hand, and more extreme precipitation events, on the other. While the relationships between temperature and precipitation extremes are difficult to quantify in a systematic manner, the above conclusion would imply that in a warmer global climate, total precipitation in the Alps would be generally reduced, but isolated events of extreme precipitation could be expected to increase significantly. This is consistent with modelling studies of Schaer *et al.* (1996), who have determined that an increase in mean temperature of 2°C over the Alps could be accompanied by large increases in extreme precipitation events (up to 30 per cent increases in the southern part of the Alpine chain).

Detailed studies such as those discussed here would be necessary elsewhere in the mountain regions of the world, primarily to characterize their current climatological characteristics, but also as a means of providing information useful to the climate modelling community. Because of the expected refinements in the physical parameterizations of climate models in coming years, and the probable increase in the spatial resolution of general circulation models (GCMs), the use of appropriate data from high-elevation sites will become of increasing importance for model initialization, verification and intercomparison purposes. The necessity of accurate projections of climate change is paramount to assessing the likely impacts of climate change on mountain biodiversity, hydrology and cryosphere, and on the numerous economic activities which take place in these regions.

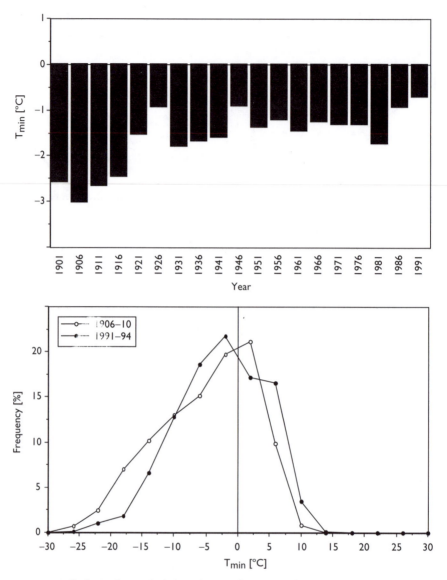

FIGURE 3.6 Shifts in the probability density function of temperature between the coldest and the warmest period of the century in Davos, Switzerland

3.6.2 Climatic variability in mountains: observations and model simulations

The IPCC (1996) Second Assessment Report documents much of the observed variations in surface temperature throughout the globe for the past century. It also provides a comparison of numerical model results of natural and forced climatic variability on a variety of time scales. Diaz and Graham (1996) have recently compared the behaviour of observed freezing-level heights in the tropics to that obtained using the results of a numerical model simulation of the atmospheric response to observed sea surface temperatures (SSTs) in the period 1970–88. The results show that the increase of tropical SSTs, which occurred in the mid-1970s, and has been documented in a number

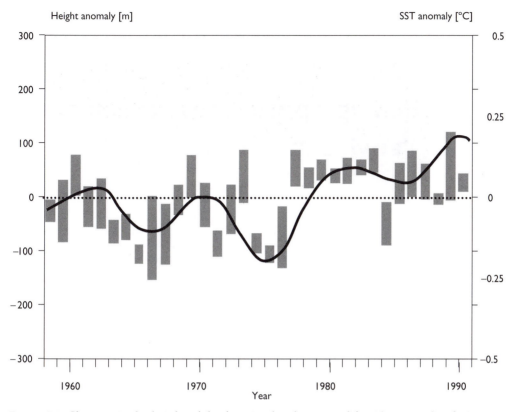

FIGURE 3.7 Changes in the height of the freezing level averaged for 10 sites in South America

of studies (e.g. Trenberth, 1990; Trenberth and Hurrell, 1994; Miller *et al.*, 1994), has led to increases in the height of the freezing-level surface in the tropics of the order of about 100 m, and consequently to warmer conditions in many high-elevation tropical ice caps, particularly in South America. This is consistent with the results of a number of studies of high-elevation tropical glaciers and icecaps (*see* Schubert, 1992, in Venezuela; Hastenrath and Kruss, 1992, in Kenya), which have shown a retreat of such features, particularly during the past few decades (Oerlemans, 1994; Thompson *et al.*, 1995).

Figure 3.7 shows the changes in freezing-level heights averaged over 10 radiosonde stations in South America. The figure also reflects the changes in global tropical SST and in tropical precipitation in the Pacific Ocean for the past few decades. A significant

upward change in the level of the 0°C isotherm took place in the mid-1970s, coincident with changes in tropical SST and precipitation (Diaz and Graham, 1996). Furthermore, the change is also consistent with a global mid-tropospheric warming as documented from a global network of radiosonde stations (Angell, 1988; Oort and Liu, 1993; IPCC, 1996).

The results of Thompson *et al.* (1995), Diaz and Graham (1996) and others also suggest that high-elevation environments in the tropics may be particularly sensitive to changes in tropical SSTs, and that prolonged El Niño-like episodes, such as those which occurred in the 1990s (Trenberth and Hoar, 1995), and decadal-scale changes, such as the mid-1970s Pacific Ocean episode, are likely to impact the hydrological and ecological balances of high-altitude zones throughout the globe.

4

Modelling approaches to assess environmental change

4.0 Chapter summary

In order to enable some insight into possible future changes of mountain and upland environments, it is often necessary to use modelling techniques. This chapter provides information on the advantages and problems related to mathematical modelling systems, how these are related to spatial and temporal scales, and the approaches which may be used to address such problems. The chapter then provides a descriptive overview of global and regional climate models, which are designed to yield information on the trends of one of the chief environmental forcing factors (climate) in mountains, as well as ecosystem models, which allow a quantification of the response of vegetation to climate change, and finally integrated assessment models. These latter models combine physical and socio-economic parameters in order to quantify the economic sensitivity of a given region to environmental change, and thus provide potentially useful information to policy and economic decision-makers.

4.1 The significance of modelling

Models are mathematical, computer-based or conceptual tools which allow us to synthesize our understanding of particular systems in the realms of physics, chemistry, biology or economics. In the field of climate research, for example, atmospheric models attempt to take into account various elements of the system which are important for its evolution. The rationale for any form of modelling is that it is usually impossible to conduct experiments with a real environmental or economic system as a whole, so that an understanding of the fundamental mechanisms governing the system emerges only through observations and modelling. Observational data provide much information on the current or past state of the system, but rarely provide any picture of its future evolution. For complex non-linear processes, which are the rule in the physical, biological or economic sciences, extrapolation to the future based on past observations cannot be considered a valid technique for most purposes. Models therefore constitute the *only predictive tool* available for investigating a particular system or process, whether it is the behaviour of air pollution in a mountain valley or the response of the global economy to stock-exchange fluctuations in a particular country. Advanced models also enable investigations of certain mechanisms which may be ultimately relevant to the understanding of a system and its prediction, such as feedbacks between the different elements which constitute the system. Furthermore, sensitivity analyses can be attempted which observation and the real world do not allow.

Models are central to our concepts of environmental physics and socio-economic issues; they cannot be dissociated from observational data, however. It is often erroneously believed that there is a cleavage between modelling approaches and experimentation or observation. Indeed, much of the scientific community today is split into two distinct groups, those working with data and those working with models. There is, however, no intrinsic reason to separate the two: data and models are by necessity complementary.

In order to interpret observational data, one needs to have formulated *a priori* some form of conceptual model to explain the information the data are providing. For example, in the case of space-borne remote sensing techniques, the information which the end-user is seeking, e.g. the temperature at the Earth's surface, is not the parameter which the satellite is measuring. A sensor on board an orbiting platform is essentially measuring radiation as emitted and reflected by the surface, and absorbed by the atmosphere as the signal is transmitted to the satellite. Such raw data need to be converted to the units of the required physical variable, i.e. temperature, on the basis of radiative transfer theory. The end-user thus requires a concept, or model, which allows a coherent interpretation of the information received by the satellite. In climate and environmental research, proxy data are those which infer a particular environmental variable from other sources of data, such as temperature from isotopes contained in lakes or ocean records, or from pollen types contained in sediments which point to past environmental conditions.

4.2 Spatial and temporal scales

O'Neill (1988) makes the point that in many environmental studies, the scale of observation determines the level of interest; a subdivision of processes into different scales is therefore useful for a particular class of problems. Furthermore, 'no single division stands out as fundamental. There certainly seems no good reason to force all problems into a single framework' (ibid, p. 32). It is additionally emphasized (ibid, p. 33) that 'the theory recommends focusing on a single level but the appropriate level should be based on the problem at hand. It is not useful to force a new problem area into the mould that was appropriate for other problems at other levels.'

Partitioning of scales is common to investigate particular sets of problems in many environmental sciences, in particular in atmospheric and climate research. Here, the spatio-temporal scales may be broadly divided into three categories, namely the micro scale, characteristic of turbulent processes, the meso scale, where regional-scale atmospheric circulations dominate, and the macro scale, associated with synoptic features and climate. Figure 4.1 illustrates a typical scaling diagram for the atmosphere, adapted from Orlanski (1975). It is seen that key processes are located along a diagonal in this space–time diagram. Processes taking place at large scales are associated with long time scales, while conversely, those operating at small scales are generally short-lived features. The fact that the range of atmospheric processes are essentially located along a diagonal enables particular combinations of spatial and temporal scales to be studied and modelled independently of the other scales. For example, mountain and valley breezes, and associated air pollution embedded within such flows, would typically require meso scale models and methods of investigation. However, climatic change and its impacts on a particular region imply processes which act at both the macro and the meso scale, and investigations of such processes would need to take into account both scales and their common interface.

The interaction between scales becomes exceedingly complex when applied to non-linear systems. Atmospheric flows are one such system which is influenced by phenomena taking place at all scales. Processes acting at the smaller scales of turbulence ultimately play a role in the evolution of the dynamic and thermodynamic state of the atmosphere through turbulent exchanges of heat, moisture and momentum. The atmospheric processes involved are thus not independent of one another simply because an arbitrary scale separation has been applied. Any investigation of a particular system therefore needs to

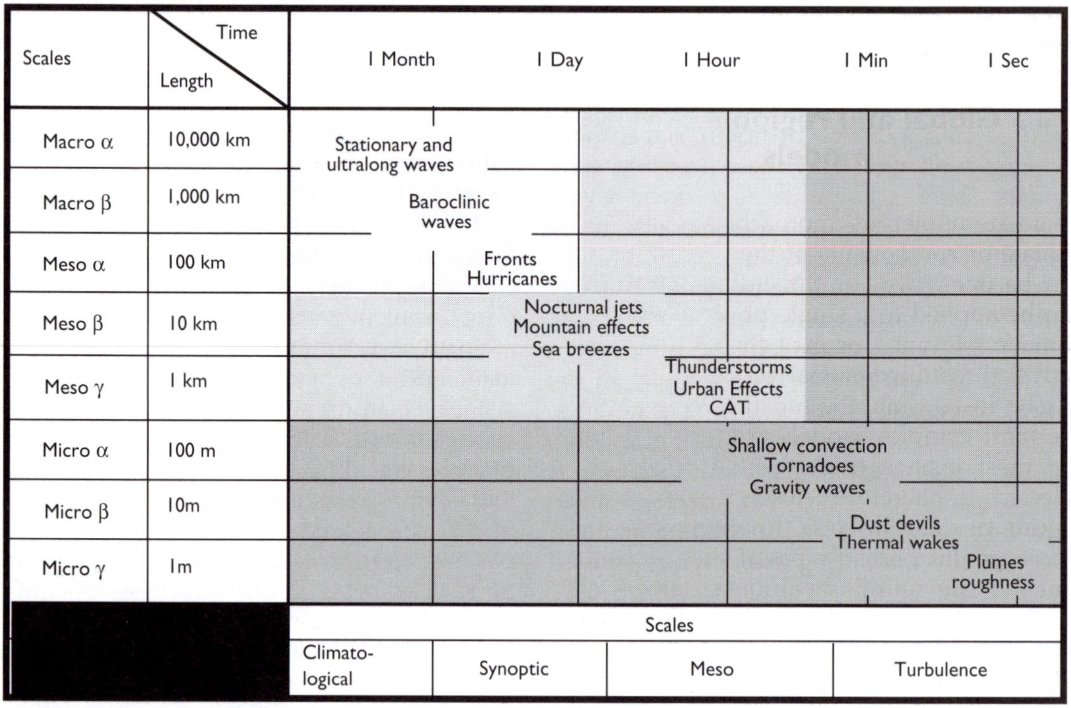

Scales	Time / Length	I Month	I Day	I Hour	I Min	I Sec
Macro α	10,000 km	Stationary and ultralong waves				
Macro β	1,000 km	Baroclinic waves				
Meso α	100 km		Fronts Hurricanes			
Meso β	10 km		Nocturnal jets Mountain effects Sea breezes			
Meso γ	1 km			Thunderstorms Urban Effects CAT		
Micro α	100 m			Shallow convection Tornadoes Gravity waves		
Micro β	10m				Dust devils Thermal wakes	
Micro γ	1m					Plumes roughness
	Scales	Climato-logical	Synoptic		Meso	Turbulence

FIGURE 4.1 Characteristic scales of the atmosphere
Source: Adapted from Orlanski (1975)

adequately determine the scales for which the objectives of the study are predominant, and to assess the likely effects of neglect of the other scales.

Small-scale features in non-linear systems are frequently responsible for apparently chaotic effects in the temporal evolution of these systems. In some instances, a long time may elapse before responses to a particular small-scale forcing become evident; when they do, however, the system may enter a totally new state. Pioneering work by Lorenz (1963) made it possible to determine the implications of chaos theory in the atmospheric sciences. Lorenz developed a nominally simple model of the climate system in which only three modes, each associated with a particular wavelength, were considered. The model suggests that weather and climate are inherently unpredictable (Lorenz, 1963, 1968, 1982; Nicolis and Prigogine, 1989), since they exhibit one of the basic properties of chaotic dynamics,

namely the sensitive dependency of the model solutions on initial conditions. Minor changes in initial conditions can result in very different solutions to the evolution of the atmospheric state after some time. It is on the basis of this consideration that the so-called 'butterfly effect' entered the popular jargon, i.e. the notion that the motion of a butterfly in one location of the globe could influence climate in another part of the world.

Despite the problems posed by chaos in dynamic systems, which renders predictability of such systems difficult, modelling has value in sensitivity experiments, whereby studies of the response of the system to a particular forcing may be analysed. Models also allow the investigation of interactions between different elements of the system in order to further the understanding of the fundamental processes and controls involved. Models are consequently a powerful means for the analysis of complex systems; they not only complement, but generally go beyond, the interpretation of observational data.

4.3 Global and regional climate models

There are numerous approaches to simulating climate or components of the system. Models can be one-, two- or three-dimensional, and can be applied to a single physical feature of climatic relevance, or may contain fully inter-active, three-dimensional processes. In order to gain insight into the functioning of climate, the most complex models are not necessarily the most explicit; the type of model to be selected is therefore closely linked to the nature of the problem to be investigated. There are three broad types of climate models, namely the quasi-one-dimensional energy balance models (EBMs), the two-dimensional zonally averaged statistical-dynamical models (SDMs), and the full three-dimensional general circulation models (GCMs), which will be briefly discussed below.

4.3.1 General circulation models (GCMs)

GCMs are essentially weather-forecasting models which operate at lower spatial resolution in order to integrate further ahead in time; they attempt to incorporate as many elements of climate as possible: not only the atmosphere, but also the oceans, the cryo-sphere and the biosphere. GCMs are among the largest and most demanding operational applications in terms of computing resources. They typically solve large sets of equations on several hundred thousand grid-points distributed in three dimensions around the globe, and these computations must be repeated 50 or more times per simulated day in order to adequately represent the temporal evolution of the system. Computer time and space requirements are thus extremely large, and GCM simulations make use of the most advanced supercomputers. In order to under-take climate simulations within a reasonable time-frame, much of the physics representing

feedbacks within the climate system needs to be parameterized, i.e. simplified in a physically coherent manner. Despite the impressive computational resources used for climate modelling, and the progress in modelling tech-niques, many uncertainties and problems remain; in particular, spatial resolution is often not appropriate for providing informa-tion on the regional scale.

General circulation models are based on the physical dynamic and thermodynamic laws governing the atmosphere. These describe the redistribution of heat, momentum and moist-ure resulting from atmospheric motion, radia-tive transfer and thermodynamics associated with phase changes of water. The governing equations are non-linear partial differential equations which require numerical methods for their solution. An overview of the govern-ing equations solved by GCMs can be found in a number of textbooks, e.g. Henderson-Sellers and McGuffie (1987) or Beniston (1997).

The numerical methods of solution consist in subdividing the atmosphere vertically into discrete layers, and horizontally either into discrete grid-points or by a finite number of mathematical functions as in spectral models. The values of the predicted variables, such as surface pressure, wind, temperature, humidity and rainfall, are integrated at each grid-point (or for each spectral function) with respect to time, in order to obtain a prediction of the future state of these variables. The time step, i.e. the interval between one set of solutions and the next, is a function of the grid interval for reasons associated with numerical stability. A GCM with a 100-km horizontal resolution and 20 vertical levels, for example, would typi-cally use a time step of 10–20 minutes. A 1-year simulation with this configuration would need to integrate the governing equations for each of the 2.5 million grid-points more than 27,000 times – hence the necessity of appropriate supercomputing resources.

Parameterization, or the simplification of the equations through physically coherent approximations, is necessary in a numerical model for a number of reasons. Certain physical processes may be acting at a scale

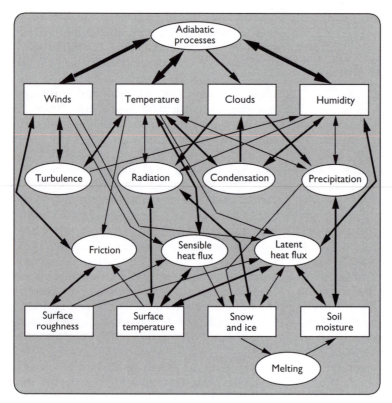

FIGURE 4.2 Typical physical parameterizations and their interaction in a general circulation model (GCM) *Source*: Adapted from Houghton (1984)

smaller than the characteristic grid interval, as in the case of turbulence. Additionally, the complete physics describing a particular phenomenon such as cloud microphysics would, if computed explicitly at each time step and at every grid-point, overload the computer resources. These processes nevertheless have relevance to the evolution of the system, so that *parametric schemes* need to be devised in order to take into account the phenomenon in a meaningful manner. A parametric scheme attempts to relate the variables at unresolved scales to those resolved at model grid-points; the quality of GCM results is directly related to the quality of the parameterized physics. Figure 4.2 illustrates the typical parameterization schemes and their interactions for the European Center for Medium-range Weather Forecasts (ECMWF) weather-forecast model. The thickness of the arrows indicates the strength of the interactions.

4.3.2 Coupled model systems

Modellers have recognized the need to incorporate whenever possible as many elements of the climate system within one model, though in practice this is an exceedingly difficult task owing to the computational resources involved and the often limited understanding of the underlying feedback mechanisms. In addition, an integrated model is attempting to simulate processes acting on vastly varying time scales, as illustrated in Figure 4.3. According to the component considered, the response time to a particular forcing can vary from a few hours to several centuries. This implies that the direct coupling of various elements of the system in an integrated climate model is by no means trivial.

In view of the significant influence of the oceans in terms of heat storage and the absorption of greenhouse gases, long-term simulations of climate require a full three-

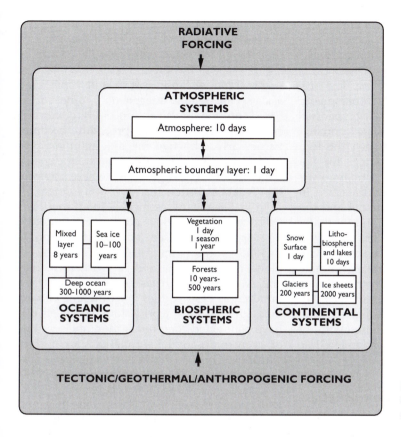

FIGURE 4.3 Characteristic time scales of various components of the climate system

dimensional ocean model, where features such as the formation of deep water are simulated. Changes in the intensity and location of deep water formation can have profound effects on the atmosphere. In the past, changes in the thermohaline circulation of the oceans have resulted in major atmospheric responses, such as the cold 'Younger–Dryas' period which affected Europe and other regions of the world after the end of the last glaciation (Broecker *et al.*, 1985; Salinger *et al.*, 1989; Street-Perrott and Perrott, 1990). Furthermore, if only SSTs are prescribed for the ocean component, climate predictability in the long term becomes questionable because features such as the quasi-periodic El-Niño/Southern Oscillation (ENSO), which exert major controls on the climate system, are not taken into account.

Until recently, ocean models underwent their development in parallel to atmospheric models, with little interaction between the two. However, recognizing the dominant role of the world ocean in climate processes, coupled ocean–atmosphere models are now emerging and have been used, for example, by the IPCC (1996) in its Second Assessment Report. Early approaches to ocean–atmosphere interactions either prescribed observed sea surface temperatures (SSTs) or specified the meridional energy transport of the oceans. This has obvious constraints for any long-term climatic simulations. Recent developments which take into account not only surface processes at the ocean–atmosphere interface, but also those acting at depth (i.e. deep ocean circulations related to the global ocean 'conveyor belt') have resulted in considerable improvements to the quality of climate model results. An oceanic GCM requires high spatial resolution to capture eddy processes, which are key features of ocean dynamic exchange, and also bottom topography and basin geometry. High-resolution ocean models are therefore at least as costly in computer time as are atmospheric GCMs.

Figure 4.3 emphasizes the significantly different response times of ocean processes compared with those of the atmosphere. The link between an ocean model and an atmospheric GCM is difficult because of the discrepancies in response times as illustrated. One approach is to synchronize coupling between the two model types at different time scales for the atmosphere and for the oceans. This involves managing a major computerized database in which the results of the ocean model and the atmospheric GCM can be stored and mutually accessed in a synchronous manner.

Coupled ocean–atmosphere systems have been the first attempt at integrated climate modelling. Further integration of other climate system component models, especially the cryosphere and the biosphere, is also necessary in order to obtain more realistic simulations of climate on decadal to secular or longer time scales.

4.3.3 Regional climate models (RCMs)

Regional-scale climatological information is important for a number of reasons, essentially related to the requirements of climatic impacts research and policy-making. In this instance, information on climate and climatic change is required at scales much finer than those at which GCMs normally operate. Furthermore, processes acting on local or regional scales often feed into the climate system, so that investigations of scale interactions as illustrated in Figure 4.3 are crucial to furthering our understanding of the climate system and the role of the smaller scales in its spatial and temporal evolution.

One technique which is increasingly used to overcome problems related to the spatial resolution of GCMs is that of 'nested modelling'. The concept behind 'nesting' procedures, which involves the coupling of models of different scales, is relatively simple. Results from a GCM for a particular region are used as initial and boundary conditions for the RCM, which operates at much higher resolution and, in most instances, with more detailed

physical parameterizations. An RCM can theoretically be used to enhance the detail of regional climatology, which would mostly be taking place at the subgrid scale of a GCM. A regional model can in this case be considered as an 'intelligent interpolator', since it is based on the physical processes governing climatic processes. Such a procedure becomes particularly attractive for mountain regions, whose complexity is unresolved by the coarse structure of a GCM grid, and where observational data are often sparse or non-existent.

In its Second Assessment Report, the IPCC (1996) has extensively used the nested modelling technique in an attempt to improve regional climate information. The method, pioneered by Giorgi and co-workers (Giorgi and Mearns, 1991), has been reported in a number of case studies by different climate modelling groups worldwide. These groups have conducted continuous monthly, seasonal or even multi-annual simulations for current climate. This includes investigations by Giorgi et al. (1993, 1994), Chen et al. (1994), Copeland et al. (1994, 1996), Bates et al. (1995) and Leung and Ghan (1995) for North America; Kanamitsu and Juang (1994), Liu et al. (1994), Hirakuchi and Giorgi (1995), Lal et al. (1995) and Sasaki et al. (1995) for Asia, notably Japan, China and monsoonal India; Cress et al. (1994, 1995), Machenhauer et al. (1994), Jones et al. (1995), Marinucci et al. (1995) and Podzun et al. (1995) for Europe; McGregor and Walsh (1993, 1994) and Walsh and McGregor (1995) for the Australasian region; and Semazzi et al. (1994) for the Sahelian zone. The horizontal RCM grid resolution in the above investigations ranges from 15 to 125 km, and the length of simulations from 1 month to 10 years.

Fewer climate change experiments have been conducted with the nested modelling technique, though McGregor and Walsh (1994), Hirakuchi and Giorgi (1995) and Jones et al. (1995), have attempted such experiments for selected parts of the world. In terms of mountain regions, the simulations undertaken by Beniston et al. (1995), Marinucci et al. (1995) and Rotach et al. (1996) in the Alpine region constitute the highest-resolution modelling experiments realized to date (*see* Section 5.5 of this volume).

When driven by analyses and observations, RCMs generally simulate a realistic structure and evolution of synoptic events. Model biases with respect to observations are in the range of a few tenths to a few degrees for temperature and 10–40 per cent for precipitation. These biases tend to decrease with increasing resolution or decreasing RCM domain size. GCM-driven RCMs tend to produce less convincing results, however, which is to be expected since lower-resolution climate models may generate unreliable results in terms of storm tracks and precipitation belts, for example. As a result, the errors of a GCM will often propagate into an RCM. As seen in Figure 4.4, however, RCMs may in some instances provide more realistic regional detail of surface climatological parameters than GCMs. This is due to the fact that RCMs capture the regional detail of forcing elements such as orography or large lakes in a far more realistic manner than GCMs. In the example illustrated in Figure 4.4, results for precipitation and temperature have been compiled by Giorgi *et al.* (1994) for an RCM operating at 50 km resolution nested into a GCM with a 300 km resolution.

The salient differences which may be observed in this illustration are directly related to the enhanced RCM resolution and to the

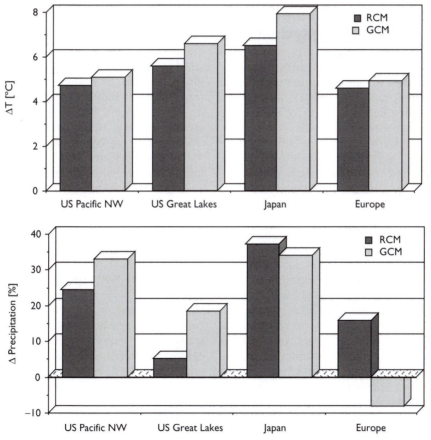

FIGURE 4.4 Comparison of general circulation model (GCM) and regional climate model (RCM) results for temperature and precipitation in selected geographical regions for simulations in an enhanced greenhouse-gas climate *Source*: Giorgi *et al.*, (1994)

improved regional detail. In this example, pre-cipitation in Europe is seen to decrease in a warmer climate according to the GCM simula-tions, while it actually increases in the RCM. This can be explained by the fact that the GCM grid poorly perceives European moun-tain regions such as the Alps, the Pyrenees or the Carpathians, while the RCM is capable of resolving the mountains and the enhancement of precipitation which they generate. In other regions, discrepancies between the two climate models are not so important, in part because either the orographic effect is not the dominant factor, or the mountain ranges involved are more adequately resolved in the GCM.

4.4 Semi-empirical methods and statistical downscaling techniques

There exist methods other than numerical mod-elling techniques for assessing environmental change in mountain regions. A number of solutions exist which help improve the quality of climate data used in impacts assessments and economic decision-making. These include mathematical techniques for downscaling from synoptic to local atmospheric scales, statistical downscaling (Gyalistras *et al.*, 1994), and use of palaeoclimatic analogues (e.g. LaMarche, 1973; Davis, 1986; Schweingruber, 1988; Graumlich, 1993; Webb *et al.*, 1993; Luckman, 1994) and geographic analogues (Beniston and Price, 1992).

Mathematical models of variation can be applied to problems of interlinks between the various scales of environmental processes. These include harmonic analysis, where high-frequency variations can be superimposed on lower frequencies. Harmonic analysis allows time-series fluctuations of different amplitudes and frequencies to be combined. Cliff and Ord (1981) have laid the theoretical groundwork for the use of harmonic analysis in scale problems, although the application to incomplete data sets remains an outstanding issue.

A further approach to the scale problem, in the form of Markov theory, can be of use as a first approximation to changes taking place

from one state to another when only the prob-abilities of transitions can be quantified. The modelling of state variations as a Markov pro-cess leads to a series of embedded Markov chains which can be analysed to assess the nature of coupling between scales. Few studies using Markov model techniques have been applied to the regionalization of environmen-tal predictions, however.

The use of fractals to reproduce patterns at smaller and smaller scales is a relatively new development which could be applied to such themes in environmental systems. Mandelbrot (1983) has suggested that all the heterogeneity observed in natural systems could be repre-sented by fractal geometry, although the application of the technique to geophysical problems has not progressed as much as expected.

Statistical downscaling is a technique fre-quently used in climate impacts assessment studies. In many instances, the scales needed for ecosystem modelling, for example, may range from an entire mountain chain down to individual plots; the data requirements are obviously very different depending on the desired end-use. When investigating the sensi-tivity of ecosystems to climate, a first short-term solution consists in the generation of analogue climate scenarios based on historical instrumental records. A comparison of the characteristics of warmer periods of the past with those of colder periods may be used to infer future changes in regional climates (e.g. Wigley *et al.*, 1980; Williams, 1980; Pittock and Salinger, 1982; Jaeger and Kellogg, 1983; Lough *et al.*, 1984). The main limitation of such scenarios is that they reflect boundary conditions for the climate system different from those expected in the future under enhanced greenhouse gas concentrations. In addition, they do not provide any information on transient climate change as expected within the course of the twenty-first century. Further-more, these scenarios are based on relatively small climatic variations, which may not be reliably extrapolated owing to the highly non-linear nature of the climate system.

A second short-term solution is based on an interpretation of the results of global climate models, in particular the results of GCM simulations. However, as has been already discussed, most climate models operate on a global scale with poor grid resolution, so that confidence in simulated changes on a regional scale is low (Wilson and Mitchell, 1987; Santer *et al.*, 1990).

The application of statistical-empirical methods represents an additional strategy in order to generate more realistic and internally consistent scenarios (Kim *et al.*, 1984; Wilks, 1989; Karl *et al.*, 1990; Wigley *et al.*, 1990; von Storch *et al.*, 1991). Compared to purely empirical or analogue scenarios, statistical scenario generation has two main advantages: first, the statistical methods allow a description of the evolution of climate variables within a given time frame, and second, the climate scenarios may be objectively linked to GCM outputs. Statistical models designed to describe local or regional shifts in climate in response to changes in global climate are generally adequate to bridge the gap between the scales typical of GCM simulations and those considered by numerous regional ecosystem studies.

The statistical downscaling method uses canonical correlation analysis (CCA) to estimate the relationships between large-scale sea-level pressure anomalies to departures of regional or local climatic parameters from their mean values. CCA is a linear regression modelling tool relating a small number of empirical orthogonal functions (EOFs; e.g. Kutzbach, 1967) to the first few EOFs of the local climate variables. The regional climatic parameters which can be generated by this method include daily mean, minimum and maximum temperature, temperature amplitude, precipitation sum, mean wind speed, relative humidity and relative sunshine duration.

Once a statistical relationship has been established, it can be validated by applying it to observed sea-level pressure variations, when these are not directly used for model estimation. The method can also be assigned to deriving scenarios of climatic change, based on GCM simulations of surface pressure changes under different conditions of greenhouse-gas forcing.

Geographical analogues represent an empirical approach which can be applied to a particular region once the required scenario data have been generated. Beniston and Price (1992) have shown that it can be of interest to investigate the regions which today experience conditions close to those which climatic change scenarios are suggesting for the future. In the Alpine region, for example, one plausible consequence of global warming may be that the Mediterranean-type climatic belts will advance northwards, thereby producing significant changes and redistribution of vegetation patterns. It is theoretically possible, within certain conceptual limits, to use geographical analogues to estimate the shifts in mountain vegetation patterns in the northern part of the Alps, in a warmer climate. This can be achieved by examining vegetation distribution patterns in the southern Alps where the climate is currently close to that predicted for the northern Alps in the future. This approach naturally assumes that vegetation and fauna can adapt to the abrupt climate change expected in the twenty-first century, that they will find appropriate pedological and environmental conditions which will ensure the continued existence of the species, and that the human barriers to migration (highways, urbanized areas) can be overcome.

4.5 Ecosystem models

Modelling is useful for the understanding of the causes and consequences of changes in species composition and to test hypotheses on the interactions between species and ecosystem processes (Guisan and Holten, 1995). In addition, it is also a valuable tool to estimate the effects of a change in environmental conditions, such as an increase in temperature. To study the impact of climatic change, large temporal and spatial scales need to be considered, which implies the use of a dynamic model. For forests, a patch dynamic or forest succession model such as a forest gap model is well suited to simulate the impact of climatic change. Climatic change can be regarded as an external disturbance to which forests respond with a secondary succession (Fischlin, 1995).

Ecologists are interested in the interactions of organisms with each other and with their environment. These are many-sided and dynamic in the sense that they are time dependent and constantly changing. In addition, feedbacks are important, and can be defined as the restoration of certain effects of a process to its source or to a preceding stage so as to strengthen or modify it. Positive feedbacks increase the effect while negative feedbacks decrease it (Braat and van Lierop, 1987). Furthermore, all living organisms are variable in their own right and in relation to other organisms through competition or predation, for example. When these characteristics are added to independent variations in environmental factors such as climate and habitat, ecological processes and systems become difficult to investigate and to control. When additionally the effects of deliberate human modification of ecological systems (e.g. agriculture and forestry) are included in ecological research, a further dimension of variability, interaction and complexity is introduced. As a result, the understanding of even relatively unmodified ecological systems is far from easy (Jeffers, 1978). Further key aspects related to the ecological approach include the wide range of time scales and the different levels of spatial scales which are involved.

A good ecological model need not attempt to reproduce every detail of the biological system (Levin, 1992). However, the objective of a model should be to determine the amount of detail which can be ignored without producing results which are in contradiction with reality (Kräuchi, 1994). Ecological models have been classified according to a wide variety of criteria. Jeffers (1978) suggests the following classification: dynamic, compartment, matrix, multivariate and optimization models.

The models discussed in this section are classified according to their spatial scale (Bugmann, 1994). The problem of selecting the optimal scale in ecological modelling is a core issue which is strongly related to the problem at hand and to the processes which may have an influence on the selection of the model parameters. An ecological system can be broken down into organizational levels. The scale of observation determines the organizational level. Higher-level forces occur slowly and appear in the model as constants, while lower-level constraints occur rapidly and appear as averages or steady-state properties. This hierarchical perspective is valuable for time and space (Jørgenson *et al.*, 1996). The spatial classification presented here ranges from global models not capable of predicting species composition, to landscape models and physiological approaches.

Micro-scale models use the knowledge of a species' natural history and physiological processes to simulate the responses of species and ecosystems to changing environmental factors (Gates, 1993). They include individual-based models and models using populations and/or functional groups. The former has its origins in the yield tables used in forest management and may be defined as a static single-species model, for example in a monotree stand. The latter type of model is used to simulate the management of single-species stands or for studies of the interactions between a few populations or functional groups of organisms. This kind of model ignores many ecological factors, often emphasizing those aspects which are relevant for forest managers such as stand structure and wood volume (Bugmann, 1994).

Physiological models are micro-scale models which simulate processes such as photosynthesis and respiration on time scales of minutes and hours (Bugmann, 1994). The most important kind of model in this category is, however, the forest gap model. Forest gap models are fairly general tools that can be used to study a variety of phenomena ranging from age structure and species composition to primary productivity and nutrient cycling of forest ecosystems. The most outstanding characteristic of these models is that they do not take into account single species. Rather, they look at a more general level of vegetation at the global, landscape and/or the ecosystem scale.

On a global scale, models are based either on a deterministic biome classification derived from the correspondence between geographic patterns of vegetation and climate (Köppen,

Box 4.1 Examples of ecological models

Micro-scale models

Botkin *et al.* (1972) presented the prototype of a forest gap model under the name JABOWA (JAnak–BOtkin–WAllis). JABOWA represents the first successful simulation of the population dynamics of trees in a mixed-species, mixed-age forest. It is a process model which simulates the succession in a forest by calculating the interrelationships among vegetation in a gap within a forest stand. Establishment, growth and death of individual trees which operate at different spatial and temporal scales were simulated on 10×10 m squares with 13 species competing for light. A site subroutine provided information about soil characteristics and calculated the number of growing degree days (Kräuchi, 1994).

Forest gap models develop a stochastic view of forest succession which assumes that a forest consists of a mosaic of patches representing the area which can be dominated by a large canopy tree and on which trees are established independently of one another. On every patch the succession of the tree species is simulated in time. To obtain forest development on the ecosystem level, the successional patterns of many independent patches are averaged (Bugmann, 1994).

In these models, tree establishment is a stochastic function of climatic (abiotic) as well as biotic factors, such as temperature, shading and the amount of leaf litter present. The growth of each tree is simulated in a deterministic manner by decreasing the maximum potential growth rate at its respective age by factors that are less than optimum. Examples of growth factors considered are the growing-season temperature, soil moisture and light availability. Tree death is determined stochastically using a function based on the assumption of a constant mortality rate throughout tree life. Most gap models include also a stress-induced mortality function that kills trees if they attain less than a certain minimum growth rate (Bugmann, 1994).

The JABOWA model was developed further and with the inclusion of additional parameters into other forest gap models such as FORET (Shugart and West, 1977), FORECE (Kienast and Kuhn, 1989), FORSUM (Kräuchi, 1994) and FORCLIM (Bugmann, 1994). All these gap models have been widely applied to a large range of forest types from the tropics to the boreal zones.

Continental- to global-scale models

The MAPSS Model (Mapped Atmosphere–Plant–Soil System) was developed by Neilson (1995) and is an ecosystem-biogeographic model which operates at regional, continental and global scales and links vegetation and water-balance processes. The conceptual framework for this approach is that vegetation distributions are, in general, constrained either by the availability of water in relation to transpirational demands or by the availability of energy for growth. The fundamental assumption under which MAPSS undertakes its computations is that the vegetation leaf area will find a maximum that just utilizes the available soil water and energy (Woodward, 1987).

The system therefore calculates the potential vegetation type and leaf area that could be supported at a site within the constraints of the abiotic climate. Both woody vegetation and grass are supported and compete for light and water.

The MAPSS Model defines the distribution of forests, savannas, grasslands and deserts with reasonable accuracy; it can also be used to model potential vegetation under altered water resources and under changed climatic conditions. Further, it can be coupled to GCMs.

Other meso- or macroscale models include BIOME2 (Haxeltine *et al.*, 1996;

Haxeltine and Prentice, 1996), DOLY (Woodward *et al.*, 1995), the Carbon Cycle Model (Goto *et al.*, 1994) and TEM (Raich *et al.*, 1991).

An example for the Alpine region

The FORCLIM Model was developed by Bugmann (1994) and is a member of the gap model family, based on the FORECE gap model (Kienast and Kuhn, 1989). The European Alps were chosen as a case study, with the aim of simulating the possible impact of climatic change on tree species composition of alpine forests along climatic gradients. Three new submodels were created; one of them included more reliable calculations of the annual sum of degree days, drought stress and winter temperature than in the previous models, which most often assumed a constant climate.

Projections for Alpine vegetation suggest that near-natural forests at mid-altitudes are likely to be resilient to anticipated changes in climate for the year 2100, whereas sites close to the summits and the dry timber-line are likely to undergo drastic changes of species composition, including forest dieback (Bugmann, 1994).

1936; Holdridge, 1947) or on functional types which are a classification of the vegetation by certain physiological aspects (Prentice *et al.*, 1992). The former estimates a change in the level of the whole biomes if an environmental factor is changed. The latter does not take the biomes as given, but they emerge through the interaction of constituent plants reacting individually to any change in their environment (ibid.). Biomes in the second approach can also be considered as emerging from physiological plant types. Therefore these models will be more capable of simulating changed climatic conditions than the first approach, which is based on correlations that may cease to apply (ibid.).

Landscape models view a region as composed of patches of ecosystems or vegetation types. Others assume the vegetation cover to be homogeneous or they concentrate on the primary productivity. However, no model of this type was designed to predict simultaneously the structure of the landscape (e.g. species composition, vegetation type) and its productivity (Bugmann, 1994).

4.6 Integrated assessment models

On the one hand it is necessary to define and assess the risks linked to climate change, ecosystem response and human adaptations; on the other hand, possible response measures must be judged in terms of their feasibility, effectiveness, cost and side effects. This requires knowledge from scientific, technical, economic and social and political fields, which must then be synthesized and communicated to those responsible for formulating policy. Integrated assessment fulfils this function, drawing from a wide variety of disciplines in order to inform policy-making and decision-making.

Whereas disciplinary research's aim is advancing basic knowledge for its intrinsic value, integrated assessments have different goals. These can be summarized as follows: first, integrated assessment coordinates exploration of possible future evolution of human and natural systems; second, it provides insights into key questions of policy-making; and lastly, it helps to uncover uncertainties and prioritize research needs to better identify solid policy options. The interdisciplinary nature of integrated assessment is essential to reaching these goals, in addition to allowing feedbacks and interactions that would otherwise be absent in individual disciplines.

The scope, geographical scale and level of detail of a particular integrated assessment model (IAM) vary with the specific needs and end-uses of the model's audience. To incorporate all the complexity of an IPCC assessment,

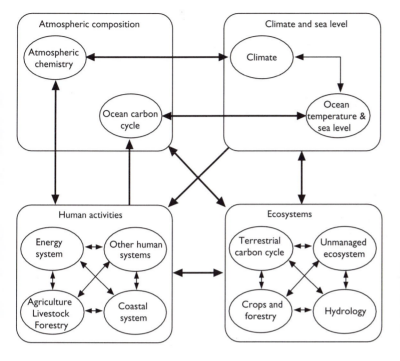

FIGURE 4.5 Elements of a full-scale integrated assessment model (IAM)
Source: Adapted from IPCC (1996)

for example, requires an attempt to represent the full range of issues associated with climate change in what is referred to as a full-scale IAM. A full-scale IAM covers a wide variety of issues that can be grouped into four main categories: human activities, atmospheric chemistry, ecosystems, and climate and sea level. Each of these categories, which interact with each other as aggregates, can then be subdivided into components which in turn interact among themselves and with other components of the main groups. Figure 4.5 illustrates a full-scale IAM's constituent parts and the numerous connections between them.

There are currently many different types of IAMs being developed according to different methodologies, by various groups of modellers, focusing on different aspects of integrated assessment. The following presents several descriptions of different types of models, according to the models' purpose or structure. Advanced models can often be used for more than one purpose.

IAMs can be grouped according to their end-use into two categories, namely policy evalua-tion and policy optimization. Policy evaluation models are used to project the consequences of different policies on ecological, economic and social systems. Rich in physical detail, policy evaluation models attempt to describe the complex, long-term dynamics of the biosphere–climate system, allowing them to point out gaps in our present understanding. The socio-economic system, in comparison, is often poorly represented. The outputs of a policy evaluation model include physical changes in emissions, concentrations, temperature, sea level and land use, as well as physical impacts of climate change on, for example, requirements for fresh water, coastal land, or ecosystems at risk.

Policy models are used to optimize key policy control variables as a function of certain predetermined policy goals. For example, in order to maximize welfare or minimize cost, this model type would be employed to optimize carbon taxes or the rate of carbon emissions control. Three types of policy optimization models exist: cost–benefit (these balance the costs and benefits of climate policy); target-based (these optimize responses given certain targets for emissions or impacts); and uncertainty-based (these provide a decision-making context under conditions of uncertainty). The shortfalls of policy optimization models are their highly aggregated representations of climate damages and the use of simplistically represented processes to resolve uncertainty. The outputs of this type of model can consist of projections of the cost of controlling greenhouse gas (GHG) emissions; economic damages resulting from climate change; control rates of GHG emissions; or the carbon tax necessary to limit GHG emissions to a specified level.

In addition to the IAM's end-use, it is possible to differentiate IAMs according to the model structure. An overall grouping thus consists either of top-down or of bottom-up approaches. Top-down IAMs are aggregate, macroeconomic models that analyse how

---

Box 4.2 Examples of integrated assessment model (IAM) results

The examples presented here serve to illustrate some concrete results obtained from actual integrated assessment models, and do not begin to cover the complete range of current models and their applications.

IMAGE2.0 (*Integrated Model to Assess the Greenhouse Effect*). This policy evaluation model, developed in the Netherlands (Alcamo, 1994), is used to asses the relative importance of environmental, land-use and socio-economic issues, e.g. climate change, land degradation, regional development and the hydrological and/or carbon cycle. Using population development and socio-economic scenarios to determine greenhouse gas (GHG) emissions from energy use and land use, it then calculates GHG atmospheric concentrations and resulting climate change. IMAGE2 subsequently calculates impacts on sea-level rise, terrestrial ecosystems and agricultural production, considering important feedbacks such as human- and climate-induced land-use changes on climate via changes in GHG concentrations and land cover (Schaeffer, 1998).

Preliminary results may confirm that one can use regional demands for land as a surrogate for measuring local land-cover changes, and that rules for land use can represent the driving forces behind land conversions. Additional results concern the effect of shifting vegetation zones on protected areas, biodiversity and nature conservation, and risk determination of specific crop production under shifting agricultural patterns. This may eventually prove to be valuable information for policy-makers faced with assessing impacts of climate change (IPCC, 1996).

DICE (*Dynamic Integrated Climate and Economy Model*) is a policy optimization model using a cost–benefit and uncertainty-based approach, developed at Yale University (Nordhaus, 1994). This globally aggregated model integrates a general equilibrium model of the global economy with the climate system, including emissions, concentrations, climate change impacts and optimal policy (ibid., 1996, p. 742). Used to balance the costs and benefits of GHG emissions, DICE has calculated that, for example, an emissions control rate (i.e. the amount of reduction in emissions relative to a baseline level) of 8.8 per cent per year is associated with a carbon tax of $4.29 per tonne of carbon (IPCC, 1996).

This model has also been used to compare discount rates. These express the time preference and risk aversion that underlie decision-making: whether to implement policy early on to avoid possible costly damages later, or to save on present investments that might be less expensive in the future. The discount rates assumed for certain political measures are critical to policy debate, as is, therefore, an IAM's ability to assess the consequences of those different assumptions.

changes in one sector affect other sectors or regions, whereas bottom-up models tend to describe the energy sector in more detail, with less emphasis on behaviour and interactions (Henderson-Sellers and McGuffie, 1997).

One can also look at the processes that an IAM incorporates and the level of detail it handles. Box 4.2 lists these processes, followed by the range of possible treatments in an IAM (a single model may have multiple treatments).

4.7 Limits and range of application of models

4.7.1 Model validation

Abstract theories and models need to be validated, in order for us to have some degree of confidence in their projections and analyses. In many cases, the limits of our understanding reduce the predictive ability of the modelling systems involved, so that the purpose of validation is not so much to prove the absolute accuracy of a model as to define the degree of uncertainty of its results and the limits of applicability. By comparing, as much as possible, model results with measurements and observations, improvements in model design or recommendations for developing new models can be suggested.

Theories and models are normally designed to address classes of similar issues. Before they can be used to answer specific questions, they must be adapted to the particular conditions of the problem to be analysed. This is achieved through the specification of initial and boundary conditions. Model initialization occurs when the state of the system is specified at the beginning of the computation, by setting each of the dependent variables to some reference value obtained from observation or hypothesis. In a GCM, for example, model initialization is not trivial, because the information used to initialize a simulation must not only be processed in a manner suited to the internal model structure, but also be consistent with the laws and relations which define the model.

Boundary conditions are also a necessary part of any model simulation. These describe constraints on the evolution of the system, or interactions between the system and its environment which are not described explicitly by the model. In climate research, many of the atmospheric data used for model initialization and verification are standardized by the World Meteorological Organization in Geneva, on the basis of a worldwide network of meteorological stations. These provide not only ground-based weather information but also upper-air soundings through the simultaneous and regular release of weather sondes around the world.

Remote sensing techniques represent a powerful method of observation and monitoring of both climatic and non-climatic environmental systems. This has been made possible in recent years through remarkable progress in satellite technology, radiation measuring instruments and optics, and computerized data post-processing and graphics. Sensors located on satellite platforms constitute the only global and repetitive source of data concerning the state and evolution of the Earth system. Remote sensing is intrinsically limited to variables that directly or indirectly affect the transfer of radiation through the environment. This means that much of the information needs to be processed, as a satellite or airborne sensor is not measuring directly the desired variable such as temperature, moisture or vegetation amount and type, but the radiative transfer through the atmosphere affected by the particular feature or variable of interest. Conversion of radiation data to the value which the end-user is seeking requires complex mathematical inversion techniques based on radiative transfer models.

It should be noted that in coming years, the capability of satellite observations will be so large that they will generate vast quantities of data, of the order of one hundred or more Gbytes of data *per day*, which is far more than any human operator can process. Even at present, only a fraction of the information received from satellites is used, owing to lack of personnel, the cost of data acquisition and the lack of appropriate software tools. Such problems are becoming increasingly crucial and will require solutions if satellite technology is to be used in an optimal manner.

Differences in observed and simulated data can help determine the improvements required in specific parts of a model's physical, biological, economic or numerical scheme. Testing the sensitivity of results to model resolution can yield information on the performance of feedback mechanisms between different elements of the system considered. Another validation method is to test separately physical subcomponents of a model directly against observational data. Validation of one isolated parametric scheme does not guarantee, however, that all interactive processes have been correctly treated. A further rigorous test of models can take the form of simulating known conditions which exist currently or have occurred in a well-documented past on Earth.

Whatever the method chosen, it should be noted that none will be entirely satisfactory or sufficient. There are no data sets allowing direct comparisons between models and reality at the spatial and temporal resolutions typically used in environmental simulations. The sparseness of available data is a general problem for any model validation, whatever the spatial scale considered.

4.7.2 Range of application of climate models

Confidence in modelling techniques may be reduced by a number of factors linked to the technique itself, a limited understanding of the biogeophysical mechanisms governing a particular system, or inadequate data for validation purposes.

Models currently suffer from insufficient spatial resolution and an incomplete description of feedback mechanisms and interactions between various elements of the system. Initialization procedures are dependent on a set of physical, biological, economic and demographic information which contain a wide range of uncertainties to which a model will be sensitive. Computational resources and mathematical algorithms, while in constant development and progress, have difficulty in

adequately modelling some of the most complex systems in existence.

Despite these often constraining caveats, however, many models have reached an advanced level of development. The results of these models, if analysed with sufficient caution, can be used to investigate the functioning and evolution of a number of environmental systems. For example, most climate models today are capable of providing an accurate representation of the broad characteristics of current climate, in terms of both average climatic conditions and seasonal to interannual variability. Such observations increase confidence in model studies of future climatic change based upon a number of socio-economic scenarios, and encourage the commitment of the scientific community to future model development.

There are some 30 GCMs which have been developed worldwide for climate research; many of these have been used in the IPCC context for assessing model capability in reproducing present-day climate, and for providing an insight into the future, greenhouse gas-enhanced climate of the twenty-first century. Much of the IPCC Second Assessment Report focusing on current GCM simulations is based on the AMIP (Atmospheric Model Intercomparison Project; Gates, 1992) experiment which has analysed model output for 30 GCMs. Other intercomparisons between GCM results have taken place in the past, such as the very detailed examination of the interactions between clouds and climate based on 19 GCMs and reported by Cess et al. (1990).

Figure 4.6 illustrates the performance of a number of independently developed, coupled ocean–atmosphere GCMs in reproducing observed summer and winter globally averaged temperatures and precipitation. Differences between one model and another are the result of differing physical parameterizations, the manner in which processes such as seasonal sea-ice and the strength of the temperature/albedo feedbacks are taken into account, and the simulation of interactions between different elements of the climate system.

While the ability of GCMs to simulate component elements of the atmosphere, ocean and

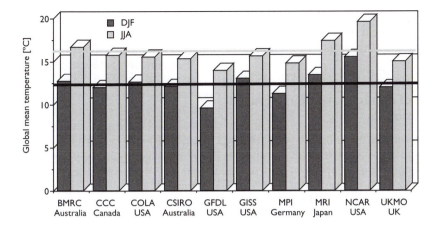

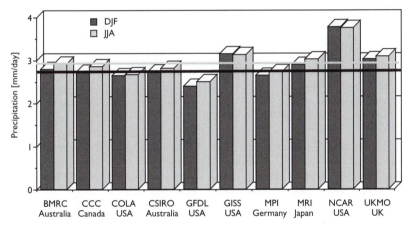

FIGURE 4.6 Globally averaged winter (DJF) and summer (JJA) temperature (upper) and precipitation (lower) simulated by 10 different general circulation models (GCMs), compared to observations (continuous lines with appropriate shading)

land surface clearly improved in the 1990s, major problems remain, in particular the radiative effect of clouds. Land-surface parametric schemes have been significantly ameliorated in recent years, with the inclusion of numerous biome types in some models, as opposed to a crude vegetation surface, and even elements which enter into the microclimate of vegetation canopies.

A reasonable simulation of the precipitation rate, as a measure of the intensity of the hydrological cycle, is especially important in coupled model systems because of possible changes in the ocean thermohaline circulation through freshwater fluxes. As seen in the lower part of Figure 4.6, the GCMs capture the broad-scale structure of global precipitation, to the degree that precipitation itself is not particularly well known on a global basis.

Because any model is by essence an approximation to reality, it cannot be used to prove a particular hypothesis. On the other hand, a model needs to be continuously tested for its robustness in the face of uncertainty. If data can be found which unequivocally demonstrate that the main assumptions, or some of them, which form part of model concept are wrong, then the model will unquestionably be shown to be wrong. Until then, however, a model will remain an acceptable theoretical framework for the investigation of a given system.

4.7.3 Range of applications of integrated assessment models

Integrated assessment modelling is a relatively young and rapidly developing field. As such, it is faced with numerous challenges and areas requiring further improvement. The interdisciplinary nature of IAMs, essential to addressing the complex issues of climate change policy, presents modelling teams with organizational problems as well. Contributions from many researchers in different disciplines, perhaps even working with different types of models, must be coordinated into one model with consistent assumptions across modules. Reconciling possible inconsistencies would improve the quality of integrated assessments. Answers to many design questions have important consequences for the model's results. These considerations, which are common to modelling in general, are compounded in an integrated modelling effort.

Integrated assessment models rely on individual scientific disciplines, and are therefore limited by the fundamental research of the underlying natural and economic sciences. Some of the research required for more complete assessments has not yet been carried out. Owing to lack of data and understanding of relative processes, for example, IAMs are only starting to investigate low-probability catastrophes – events that are of considerable concern to policy-makers.

Numerous uncertainties remain to be resolved in integrated assessment of climate change. Among these are the sensitivity of the climate system to changes in GHG concentrations, the physical and economic impacts of climate change, non-market impacts, changes in demographics, the choice of discount rates, and the assumptions regarding cost, availability and diffusion of technologies. IAMs can deal with uncertainty using either sensitivity analyses on model inputs, or formal risk or decision analyses assigning probabilities to

model inputs. These methods allow the screening and prioritizing of uncertainties but do not, however, eliminate them.

A significant problem with IAMs concerns their treatment of developing and transition economies. Countries with population growth and economic development that do not correspond to developed countries' patterns are poorly represented in integrated assessment models. An urgent need exists to represent these countries' situations more realistically, especially regarding land-use types and controls, the informal economic sector, institutional barriers, hierarchical decision structures, market imperfections and, most importantly, population growth. Without more detailed and accurate information in these areas, outputs of global assessments tend to be rather biased.

Progress is only recently being made in representing climate change and evaluating its impacts more credibly in IAMs. Most models are based on equilibrium climate models, a situation that is not necessarily realistic. The use of transient climate models (i.e. where GHG concentrations increase progressively with time) in IAMs would allow projections of transient responses of impacted systems, for example successional paths of ecosystems and adaptation capabilities of agrosystems (e.g. Schaeffer, 1998). Many uncertainties surround this newer approach, as it is still in its infancy. In terms of evaluating impacts, information needed to effectively project impacts is not yet determined, and valuation methods are still being developed.

It has been suggested that impacts from climate change and climate change policies can be correctly evaluated only when put into a context where perception and interpretation of climate are also represented in an IAM. Since measures dealing with reducing climate change impacts can be successful only when a society accepts them, the effectiveness of impact assessment may well be improved by including these societal processes (Hulme, 1997).

Finally, an integrated assessment model must produce information that is relevant and understandable for policy-makers and the general public. Examples of relevant information not yet included in IAMs are income distribution, unemployment, non-market impacts, etc. IAMs tend to represent social forces weakly, with a bias towards the natural sciences. Also, delivering assessment information in an understandable, concise and usable form is essential to applying it to policy formulation. Models that generate globally aggregated information do not consider that climate impact decision-making is usually carried out at a national level (Nordhaus and Yang, 1996).

Possible future changes in mountain environments

5.0 Chapter summary

This chapter provides up-to-date information on changes which can be expected in mountain environments in coming decades, based on the modelling techniques discussed in the previous chapter, and the baseline conditions estimated from palaeoenvironmental studies as outlined in Chapter 3. A distinction is made between natural forcings, such as ENSO and the North Atlantic Oscillation, and anthropogenic forcings. In this latter case, the causal mechanisms leading to environmental stress are discussed, with a focus on demography, economic growth and resource use in both the developing and the industrialized worlds. The likely shifts in environmental conditions resulting from pollution, deforestation, desertification and climatic change are discussed and, wherever possible, examples for specific mountain regions are given.

5.1 Natural forcings

It is exceedingly difficult to estimate the manner in which mountain environments are likely to change in the future. Multiple stress factors will interact to shape the evolution of soils, vegetation, hydrology and climate. It is probable that the response of mountain environments will be highly non-linear, and that the combined effects of multiple stresses will be larger than the sum of the individual stresses.

In addition, environmental change in mountains will be largely determined by forcings that are external to the mountains, i.e. on geographical scales much larger than that of the largest mountain chains. These forcings will be both natural and anthropogenic; while this chapter will focus on the latter, natural forcings are numerous and warrant an overview in order to highlight the fact that the disaggregation of natural versus human-induced forcings is far from trivial.

5.1.1 El Niño/Southern Oscillation

One example of external forcing which has far-reaching implications for numerous regions, including mountains, is the so-called El Niño/Southern Oscillation (ENSO), which represents an interplay of coupled ocean–atmosphere phenomena. El Niño is the warm phase of ENSO, whereby a weakening of the prevailing easterly trade winds in the equatorial Pacific allows the eastward propagation of warm waters which normally accumulate to the west of the Pacific basin. Associated areas of deep convection 'migrate' with the propagation of the warm waters, which are the principal energy source for convection. The area of anomalously warm waters at the peak of an El Niño episode can reach 30 million km^2, roughly three times the size of Canada, and as a consequence the sensible and latent heat exchange at the ocean–atmosphere interface is sufficient to perturb climatic patterns globally. This perturbation occurs in three simultaneous steps: vertical transfer of energy, heat and moisture through the deep convection; horizontal propagation through atmospheric flows at high elevations; and, in time, an 'overflow' into the mid-latitude synoptic systems which can reinforce or weaken surface pressure patterns and deflect the jet streams from their usual trajectories. Mountain regions are affected by ENSO events, through extremes either of drought or floods, because of the general reversal of normal precipitation patterns. Areas such as South-East Asia (the Philippines, Indonesia and Papua New Guinea) or Australia frequently experience unusually dry conditions during El Niño. Normally arid regions of Latin America (in particular Peru and northern Chile) as well as North America (Mexico; California) can be subjected to extreme precipitation events, with associated mudslides and landslides. Areas of the globe other than the Pacific Rim are also affected by ENSO, such as Southern and Eastern Africa, the Indian subcontinent, Brazil and Argentina. Figure 5.1 illustrates the regions typically affected by anomalous weather patterns, following CLIVAR (1997).

The cold phase of ENSO, commonly referred to as La Niña, occurs sometimes (but not always) at the end of an El Niño event. Anomalously cold waters invade the tropical Pacific region, and the strength of La Niña can in some instances reverse the previously discussed anomaly patterns, i.e. by enhancing respective precipitation or drought patterns where these normally occur.

There is speculation (Wolter and Timlin, 1998) that global warming may enhance the frequency and intensity of ENSO events. This is based on the observation that since the early 1970s, the return period of the warm phase of ENSO has substantially decreased, as shown in Figure 5.2. The figure shows, in terms of peak temperature anomalies, not only the increasing trend and frequency, but also the reduction in the frequency of the cold La Niña phase of ENSO. Following the 1992–93 event, the waters of the equatorial Pacific did not return to normal conditions for a number of years. Although this may be unrelated to observed warming and could be explained by chaotic behaviour of the highly non-linear ocean–atmosphere coupling, if a warmer climate were to be accompanied by increases in El Niño events, then the consequences for

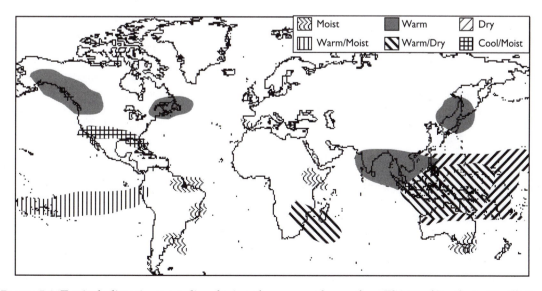

FIGURE 5.1 Typical climatic anomalies during the warm phase of an El Niño/Southern Oscillation (ENSO) event

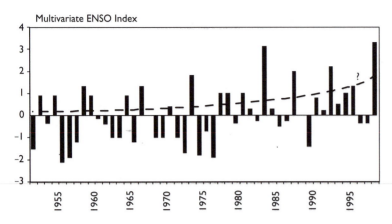

FIGURE 5.2 Time series of peak warming during El Niño/ Southern Oscillation (ENSO) events since 1950

mountain environments and economies could be dramatic. Natural catastrophes related to strong ENSO events lead to immense hardship in terms of loss of life and displacement of entire populations, and total billions of dollars in losses as a result of floods, droughts, crop failures and forest fires. For example, the 1982–83 event reached some $8 billion world-wide, according to Swiss Re (1997).

The 1997–98 El Niño was the most intense event of the twentieth century, resulting in massive forest fires from Indonesia to Australia, catastrophic flooding in southern and central California, Peru and Chile, and crop failures in the Philippines, Indonesia (rice), Brazil and southern Africa (millet; tropical fruit). It coincides with the warmest period of the century, lending some credibility to the speculation that there may be a link between global warming and increases in El Niño events. The recent spate of El Niño episodes has contributed significantly to the increase in global mean temperatures in the 1990s.

5.1.2 The North Atlantic Oscillation

The North Atlantic Oscillation (NAO) was seen in Chapter 3 to strongly influence precipitation and temperature patterns in both the eastern third of North America and Western and Northern Europe. As in the case of increasing ENSO events, there is the possibility that changes in the behaviour of NAO may be related to the observed warming, particularly since the early 1980s. The current state of

knowledge of the fundamental mechanisms underlying NAO is still too tenuous for us to come to any definite conclusions. However, if the patterns observed in the 1980s and 1990s were to become more common in a warmer climate, the associated precipitation and temperature patterns could lead to significant changes in the mountain vegetation, hydrology and the cryosphere, particularly in Europe (Beniston *et al.*, 1994). Positive NAO index values are linked to anomalously low precipitation and anomalously mild temperatures, particularly from late autumn to early spring, in Southern and Central Europe (including the Alps and the Carpathians). The reverse is true for the mountains of Norway, where a positive mass balance for a number of glaciers there led to surges in the late 1980s and early 1990s, while at the same time glaciers in the Alps were experiencing a generalized retreat. Figure 5.3 illustrates that, without the additional influence of persistent and positive NAO anomalies in the last 20 years of the twentieth century, global mean temperature increases would have been substantially lower (Hurrell and van Loon, 1997). An abrupt change in the NAO occurred in the winter of 1995–96, reversing the patterns described above.

5.1.3 Volcanic activity

Volcanic activity can substantially modify mountain landscapes and environments in two manners. The direct effect of volcanoes

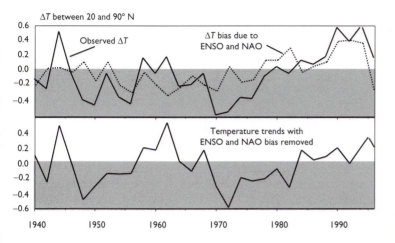

FIGURE 5.3 Observed global mean temperature anomalies, and the respective bias of El Niño/Southern Oscillation (ENSO) and North Atlantic Oscillation (NAO) in the record

occurs on a local or regional scale through the deposition of material in lava flows and fallout of ash and other potentially nutrient-rich elements, which can transform the mineral and organic content of mountain soils. The indirect influence of volcanic activity has far-reaching effects on global climate, through the emission of aerosols into the high atmosphere; these aerosols reside for periods of several months to several years and act to reduce the incident solar radiation intercepted at the Earth's surface. This reduction of available energy can perturb temperature and precipitation patterns and their impacts on numerous regions of the globe, including mountains. Prediction of volcanic activity is currently impossible, and while there seem to be periods when this activity is more intense than at other times, there is no way to ascertain at present how this may change in the next decades to centuries.

5.2 Causal mechanisms of anthropogenic pressures on the environment

5.2.1 Demography

Natural forcings are certainly important on various spatial and temporal scales, but a remarkable feature since the late nineteenth century is the increasing interference of humankind with natural environmental systems. Population growth has undergone a fivefold increase since the beginning of the twentieth century; world population as it enters the twenty-first century stands at more than 6 billion. Demography is indirectly responsible for perturbations to the environment in two ways: through increasing demands on natural resources (energy, agriculture, water), and through economic growth to maintain or augment a population's standard of living. There are thus subtle links between the rate of population growth, the rate of economic growth, and environmental pressures. It is necessary to distinguish between these factors in developing countries and in the industrialized world, because a large population in a developing country may in fact be consuming less energy and fewer natural resources than a smaller population in an industrialized country. It is thus not just a country's population but also its level of development that determines the extent to which the natural environment is put under pressure. For example, according to Table 1.1, the total annual energy use in Ethiopia is about 45 terajoules (i.e. 45×10^{12} joules), while that of Switzerland is 952 terajoules, i.e. more than 20 times the consumption of Ethiopia for a population which is an eighth the size.

In recent years, there has been a decrease in birth and fertility rates, and population growth rates have peaked everywhere except in Africa. In absolute terms, however, there is a rapid increase in global population, and this is likely to continue throughout the twenty-first century. Although the number of children born per woman decreased in the latter part of the twentieth century, the number of women of child-bearing age increased substantially. As a result, there is a large imbalance in the age structure of countries in the developing world, with a disproportionate number of young persons. This is a direct consequence of the rapid growth of these populations in the latter half of the twentieth century. In many industrialized countries, on the contrary, fertility rates are often not sufficient to maintain the indigenous population at current levels; as a consequence, the age structure shows an increasing trend towards more elderly age groups.

On a national scale, the age structure of a population is crucial to its economic development; a rapidly expanding population in a developing country may negate its efforts at raising living standards. On a global scale, rapid growth has wide-reaching implications for regional and global environmental issues, because of the requirements for food, water, housing, energy and employment. According to WRI (1996), the industrialized countries have been consistently forecast to show only very small increases in population in coming decades, primarily because of the very low total fertility rates of many European countries and North America. The lowest fertility rates occur today in Italy (1.3 children per woman of child-bearing age), Spain (1.5) and Sweden (1.9). The replacement-level fertility is 2.1 children per woman, and today most Western European countries as well as the United States (1.9) are below this level. Modest increases in population in many Western European countries in recent years are more the result of immigration than of indigenous population growth. In the developing countries, the current population growth rates will slow down during the twenty-first century, but nonetheless there will be significant increases in population in some Asian countries (particularly Bangladesh, India, Iran and Pakistan) and many African nations. As a result, there will be major differences compared to today

Table 5.1 IPCC scenarios for world population growth by the year 2100 (in billions)

IS92A	IS92B	IS92C	IS92D	IS92E	IS92F
11.3	11.3	6.4	6.4	11.3	17.6

Note: World population in 1990 was 5.25 billion

in the percentage of world population living in the different continents, with the largest increase between 1987 and 2025 projected for Africa (1987: 11.8 per cent of the world's population; 2025: 18.7 per cent).

The IPCC (1992) developed a number of socio-economic scenarios to estimate the rates of increase of greenhouse gases in the atmosphere in order to assess the sensitivity of the climate system to this increase. Labelled IS92A to IS92F, the six scenarios combine demographic projections and economic growth perspectives based on a number of criteria. Table 5.1 gives the global projections to the year 2100. The very modest population rise assumed in two of these scenarios would require significant changes in social patterns in different parts of the world. A case in point is the fact that women, who account for slightly more than half the world's population, do most of the domestic work and child care. They also undertake much of the labour associated with growing food, drawing and carrying water, and gathering fuel-wood. Women who have access to education and paid labour tend to have fewer children; the IPCC (1992) has assumed in the IS92C and IS92D scenarios that such access could become more widespread in the future. As a result, women would be in a position to decide in their own right on whether to bear children or not, thereby limiting population growth very substantially, and in a far more efficient manner than childbirth reduction programmes imposed by governments. However, for such a revolution to take place, awareness that rapid population increases undermine a country's wealth and environmental quality needs to be raised. In the IS92F scenario, the assumption is that the gap between poor and rich nations will widen even more rapidly than currently; in this case, children in the poorest countries are seen as a form of 'capital' in the economic sense, from the family level up to the national level. In this case, world population will rise to almost 18 billion, essentially in the developing world.

When considering increases in world population, it is also necessary to take into account not only the age structure, but also the geographical differentiation between urban and rural populations. Urban areas are generally more resource-intensive than rural areas; the average wealth is higher (even if there are large social discrepancies within cities) than in the countryside, particularly in poorer nations. Consequently, the use of fossil fuels for heating or cooling, and for transportation, is much more important than in rural regions. In addition, cities are not self-sustainable, because they need to import most of their food, water and energy resources from the outside, both at the national and the international levels. The flux of resources is poorly managed, so that most major cities are significant sources of air and water pollution, through industry, traffic, and domestic and industrial waste. This pollution ultimately reaches the environment at large through atmospheric and hydrological transport and diffusion processes. The groundwater table and soils are also put under stress from long-lasting contaminants such as heavy metals produced by urban industrial sites. The situation is particularly critical in large urban centres of the developing world, where environmental legislation is either lacking or not implemented, and where commodities that are taken for granted in the industrialized world, such as safe drinking water, are practically non-existent. In such a context, the perception is that there are more urgent matters to deal with in terms of sanitation and poverty than in terms of environmental protection.

UNEP (1993) shows that world urban population in 1990 was 41 per cent of the world total. This figure can be broken down into developed and developing countries, where the respective proportion of urban dwellers was 72 per cent and 33 per cent. In the twenty-first century, the largest cities of the world will be in the developing countries and no longer the monopoly of countries such as the United States or Japan. Already, cities such as Mexico City, Cairo or Jakarta have populations that exceed or compare in size with New York, Los Angeles or Tokyo. Of the 5 billion urban dwellers projected for the year 2025, representing a probable 65 per cent of the world's inhabitants, over 4 billion will be in the developing world and 1 billion in the more industrialized countries. This will not be without numerous and far-reaching implications for environmental degradation and the propagation of the negative impacts of pollution transported towards mountain regions.

In certain mountain regions, principally in Europe and to a lesser extent in North America, urbanization is transforming the landscape. Mountain resorts that were originally agricultural communities are now small towns that are capable of providing accommodation and commodities to several thousand persons. The presence of constructed zones in fragile environments is accompanied by a range of negative side-products resulting from tourism, such as pressures on water resources, domestic waste and road infrastructure, which encroach upon the landscape. There is a growing awareness of such problems, and it is difficult to assess how urbanization in mountain regions will develop in the future. It is possible that there will be a levelling-off of construction in the industrialized countries, with an increase in the developing world, as population pressures generate demand for housing and infrastructure in mountainous countries such as Bolivia, Mexico or Nepal.

5.2.2 Economic growth

The level of economic development of a country determines its dependency on natural resources such as water and energy. In a highly liberalized economy, where economic growth stimulates development patterns, increasing growth is generally accompanied by increasing demand for goods and services, and therefore greater environmental pressures. Economic growth is determined by maximizing the flow of resources and energy by means of enhanced consumption per capita, increasing population, or a combination of both. It is often quantified in terms of a country's gross national product (GNP); this indicator is a poor measure of human welfare and environmental quality, however. The GNP does not quantify the negative impacts of producing goods and services, for example in terms of air, water and soil pollution or in terms of human health. It does not include any criterion related to environmental degradation and resource depletion upon which national and supranational economies depend. The GNP also masks problems of social and economic equity, in terms of unemployment, and distribution of wealth.

An additional cleavage between many schools of economic and environmental thought is related to internalization or externalization of costs. The costs of manufacturing a particular product (raw materials, labour, distribution into the marketplace, for example) are referred to as internal costs. There are also a number of external costs which enter into a manufactured product but which the manufacturer and the consumer do not pay for, at least directly. These include the extraction and processing of limited natural resources for manufacturing a product, the use of non-renewable energy in the production process, and the generation of waste and other forms of pollution, which as well as having other negative effects can affect human health. These external costs are borne by society at large and not by the manufacturer and the individual consumer. In this context, the environment is seen by many pro-growth economists to have no intrinsic economic value, and therefore, they believe, no effort to conserve natural resources and to prevent environmental degradation needs to be made. If current economic patterns continue into the future, there could rapidly be adverse consequences for environmental quality as the basic resources are depleted at an ever-faster rate. These global considerations would also have repercussions for mountain regions, particularly in terms of the exploitation of mineral, wood and energy resources.

Poverty is a factor linked to economic growth, which can have direct and indirect consequences for the environment and accelerate the spiral of poverty, both in the industrialized world and in the developing countries. Poor people are often exposed to greater health and environmental risks, and in countries with growing populations these risks will increase in the future. In terms of the distribution of wealth, the gap between the rich and the poor (both within individual nations and between rich and poor countries) has widened steadily since the 1960s (Miller, 1996). Eighty-five per cent of the world's economic wealth is in the hands of 20 per cent of the world's population, and this gap is likely to widen in the future, because 95 per cent of the projected increase in world population will take place in the developing countries. Other signs of social inequity include child labour, which currently affects an estimated 200 million children, a figure which is projected to double by 2010. Unemployment is also a by-product of free-market economies, in which human resources are considered to be factors that can limit profit margins. As in the case of external costs, those related to unemployment are borne by the entire community and not by those sectors of the economy which contribute to this situation. A further factor which needs to be taken into account is the increasing number of 'economic refugees', i.e. the migration of large numbers of persons from poor regions to richer nations. Such fluxes of population are likely to increase in the future, particularly through illegal immigration. Most governments are today ill-equipped in legislative terms to deal with this type of situation. The political and economic tensions that are raised by refugees could lead to conflictual situations in many regions.

While such considerations may appear to be far removed from the theme of environmental change in mountain regions, aspects such as poverty or migration do have indirect effects on the environment. Although persons concerned by inequity in one form or another have far more urgent affairs to attend to than environmental quality, the impoverishment of a population will inevitably lead to environmental degradation as resources are used in order to survive.

TABLE 5.2 IPCC scenarios for world economic growth by the year 2100 (% per year)

IS92A	IS92B	IS92C	IS92D	IS92E	IS92F
2.3	2.3	1.2	2.0	3.0	2.3

The IPCC (1992; *see* Table 5.1) developed, in the context of its socio-economic scenarios IS92A–F, projections for economic growth which are provided in Table 5.2. The figures given in Table 5.2 span a threefold range in terms of global economic growth. The differences are linked to different assumptions concerning energy consumption trends and pricing structures for oil and gas, which could lead to a switchover to renewable energy sources. Other assumptions include differential growth rates between currently industrialized countries and developing countries, technological improvements to production and energy consumption, and the rate of penetration of 'environmentally friendly' technology into the marketplace. The figures in Table 5.2 should be viewed with caution, as they reflect a number of uncertainties that are inherent to the prediction of the highly non-linear nature of economics. In addition, it is impossible to foresee several years or decades in advance a number of political upheavals, such as the collapse of the Soviet Union and the centrally planned economies of Central and Eastern Europe in the early 1990s. These political changes have had far-reaching consequences for regional and global economies, and there are likely to be a variety of unexpected changes of this sort during the twenty-first century.

5.3 Environmental pollution

As in many other cases of environmental stress, exogenous factors will contribute to the overall environmental impacts in mountains and uplands. Air pollution is one such stress factor which can have a number of ecological and social consequences, because it is so readily transported by atmospheric circula-

tions from its origin (the emission source) to its receptor area. Chemical compounds in the atmosphere can be dispersed over large distances, according to the type of pollutant, its chemical reactivity and the dominant airflows in which it is embedded. As a result, particulate matter or acidic compounds are found in remote mountain regions which are removed from their source regions by hundreds or thousands of kilometres. This section will provide a brief overview of the principal anthropogenic pollutants, other than the greenhouse gases, which will be discussed in a later section.

As a result of a variety of human activities in an increasingly populated world, a large number of toxic substances are emitted into the atmosphere. Among the chemicals that pose problems are pesticides, polycyclic aromatic hydrocarbons (PAHs), dioxins, and volatile organic compounds (VOCs) such as benzene and carbon tetrachloride. Because of the enormous variety of toxic pollutants present in the air, it is a challenging task to determine the potential ecological and human health hazards. It can be safely concluded, however, that the compounded effect of numerous chemical species is likely to be more severe than the sum of the impacts of individual chemical species.

In both the industrialized and the developing world, air pollution is associated with combustion, industry and traffic. Smoke and a variety of chemical species combine to form smogs, which have prevailed since the beginning of the industrial area in Europe and North America, and are currently also encountered in many parts of the developing world. In most OECD countries today, however, severe air quality problems significantly diminished in the latter part of the twentieth century as a result of changing fuel-use patterns as well as the implementation of effective emission controls.

The major problem worldwide is currently related to pollution from traffic. Combustion engines in private vehicles and trucks emit pollutants such as carbon monoxide (CO), nitrogen oxides (NO_x), volatile organic compounds (VOCs), particulates and hydrocarbons (which are produced essentially from unburned fuels). These compounds have an increasing impact on urban air quality and on regions situated downstream of the prevailing atmospheric flows. Under certain conditions, chemical transformations will convert the nitric oxide (NO) to nitrogen dioxide (NO_2). NO and NO_2 are produced primarily by road traffic, which is heaviest in urban areas, and accounts for what is generically referred to as 'urban smog'. Nitrogen dioxide has a variety of environmental and health impacts, in particular in terms of respiratory ailments.

NO_2 is one of the precursor gases of ozone (O_3), which is a highly corrosive and toxic gas that damages plants and leads to respiratory and ocular problems at sufficiently high concentrations. Ground-level ozone is not emitted directly into the atmosphere, but is a secondary pollutant produced by reaction between NO_2, hydrocarbons and sunlight. Whereas NO_2 participates in the formation of ozone, NO destroys ozone to form oxygen (O_2) and NO_2. For this reason, ozone levels are not as high in urban areas (where high levels of NO are emitted from vehicles) as in rural areas. As the nitrogen oxides and hydrocarbons are transported out of urban areas, the ozone-destroying NO is oxidized to NO_2, which participates in ozone formation. Ozone can therefore be present in mountain areas located at considerable distances from the source regions of the ozone precursors. Ozone is considered to be one of the factors responsible for forest dieback in the European Alps, for example, where damage to trees may be linked to heavy industrial sites at the boundaries of the mountains, as in Lombardy (in the north Italian plain). As always with environmental threats, the large time lag between load and full reaction, where ecosystems and soils are involved, means that pollution has probably still not reached its peak impact.

Grassl (1994) has shown that there has been an upward trend in tropospheric ozone concentrations at Alpine sites in Germany (Zugspitze, at 2950 m altitude) since measurements began in the late 1960s. In contrast, ozone levels at the nearby urban site of Garmisch-Partenkirchen have not risen

(Schneider, 1992), typically because of the ozone-destroying chemical reactions taking place in an urban environment with a high density of motorized traffic. Because sunlight is a prerequisite for the physico-chemical transformation of NO and NO_2 into O_3, high levels of ozone are generally observed during periods of persistently warm weather, and in locations where poor ventilation has allowed concentrations of hydrocarbons and nitrogen oxides to reach critical levels. Because of the time required for chemical processing, ozone formation tends to take place downwind of pollution centres. The resulting ozone pollution or 'summertime smog' may persist for several days and be transported over long distances.

Sulphur contained in coal, which is used in many industrial processes worldwide (energy supply, smelting, etc.), is released into the atmosphere in the form of sulphur dioxide (SO_2). This can then be dissolved in liquid water droplets in clouds and fog to form sulphuric acid (H_2SO_4), which is then transported in atmospheric flows and ultimately falls to the surface in a process known as acid deposition or acid rain, with a number of adverse environmental consequences. Acidification of soils damages vegetation and releases heavy metals such as aluminium, copper or mercury into hydrological systems; these toxic metals are absorbed by fish and other aquatic species, with serious consequences for the food chain. The interactions between living organisms and the chemistry of their aquatic habitats are extremely complex. If the number of one species or group of species changes in response to acidification, then the ecosystem of the entire water body is likely to be affected through predator–prey relationships. Terrestrial animals dependent on aquatic ecosystems for their food supply, such as birds, are also affected.

In the European Alps, for example, the consequences of acid precipitation may be exacerbated in many mountain regions by the fact that precipitation increases with altitude. Smidt (1991) has shown that, for two Austrian alpine sites, contamination of precipitation with ions resulting from pollution decreases with altitude, but deposition increases because of enhanced precipitation with height. The deposition of hydrogen ions thus strongly increases with height. Since the concentration of basic anions and cations in precipitation is rather uniform over Central Europe, the Alps are subject to as much acid deposition as other areas because of the orographic influence on precipitation, despite the fact the Alps are not in themselves a major source of sulphate-based pollutants.

Heavy metals are injected into the atmosphere through combustion processes, for example from unleaded vehicle fuels, metal processing industries and waste incineration. Lead, cadmium and mercury are highly toxic substances which enter the food chain through wet and dry deposition on cultivated areas; they can then be directly absorbed by humans in fruit or vegetables, or indirectly through meat from animals which have themselves ingested and concentrated these metals. Heavy metals affect the central nervous system in humans.

Radionuclides are another form of long-distance pollutants which can affect mountain regions. The health and ecological hazards of minor quantities of certain radioactive elements can be considerable, as a result of exposure to the element over time, or because of the intensity of radiation from the substance, or a combination of both. A case in point is the accident of the Chernobyl nuclear power station which took place in April 1986. The release of highly radioactive material into the atmosphere, and the particular configuration of high- and low-pressure cells over Europe at that time, resulted in strong levels of caesium-137 (^{137}Cs) in the southern part of the Alps and Austria. After having been airborne in its trajectory from the Ukraine, over northern Scandinavia, the British Isles and southern France, a rainy episode washed out some of the material from the radioactive cloud and deposited ^{137}Cs to the surface in many regions, as illustrated in Figure 5.4. Caesium-137 has a half-life of roughly 28 years, implying that its potentially adverse effects may be felt into the 2020s and beyond.

Stratospheric ozone depletion caused by the catalytic destruction of ozone molecules can indirectly affect mountain regions, particularly in the extreme southern part of the Andes, but increasingly in high northern latitudes too. This is because the resulting rise in levels of

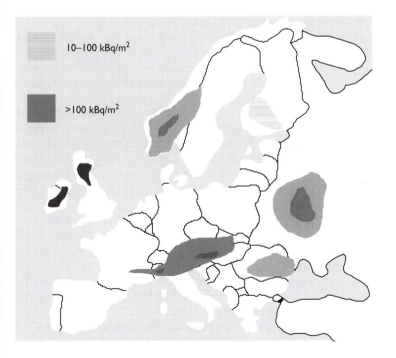

10–100 kBq/m^2

>100 kBq/m^2

Figure 5.4 Concentrations of caesium-137 in Europe following the 1986 Chernobyl nuclear accident. Note the impact on Scandinavian and Alpine zones
Source: Stanners and Bourdeau (1995)

biologically effective ultraviolet radiation (UV-B) would affect photochemistry in the troposphere as well as photosynthesis in high-altitude mountain areas.

Future trends in pollution and its long-distance transport are difficult to assess. On the one hand, rising population and increasing industrialization of the planet have the potential for generating high levels of pollution. On the other hand, the awareness that pollution is detrimental to human health and the environment has led to policies aimed at controlling emissions, with the corollary that air quality has improved in many regions. Regional aspects of air pollution may continue to dominate in some regions; an example is heavy truck traffic linking Northern and Southern Europe across the Alps (particularly the Gotthard highway in Switzerland and the Brenner highway in Austria). This is currently a cause of concern and political discussion between the transit countries and the rest of the European Union, which is a major promoter of truck transportation. The developing countries are lagging behind in terms of pollution abatement, however, and unless the problems of air quality in these countries are taken seriously into consideration, the situation is likely to deteriorate over time. Thus the impacts linked to the deposition of a number of chemical compounds in many mountains and uplands are likely to continue to be an issue in coming decades.

5.4 Land-use changes

5.4.1 Deforestation

The overexploitation of forest resources over centuries, mainly in the face of population pressures and the consequent need to provide land on which to grow food, has led to a situation where many forested areas of the world are now in rapid retreat. As people seek land to cultivate, and wood for fuel, industry and construction, they turn to tropical forests, which are being destroyed at an unprecedented rate. This destruction has wide-reaching consequences for economic, social and ecological

factors; it is the world's poorest people who are the most severely affected.

Most cultivated and inhabited lands have been claimed from previously forested areas. For centuries, forest destruction has been at the expense of dwindling species habitat and, in some cases, species extinction. Since pre-agricultural times, about 20 per cent of global forest resources have been lost, mostly in the temperate regions of Europe, Asia and North America. The Himalayan watershed covering northern India, Nepal, and Bangladesh had lost 40 per cent of its forest by 1980. The United States cleared most of its forests in the nineteenth century and is still felling trees. Costa Rica has lost a third of its forests, loses 60,000 hectares a year and was projected to have none by 2000 except in highly fragmented biosphere reserves. The coniferous taiga is one of the last remaining forest areas in Siberia. In recent years, 25 million tonnes of Siberian timber has been transported westwards annually as part of the new Russian economic expansion. In addition, deliberate fires (to create pasture) and accidental fires consume 1 million hectares of Siberian forest every year.

In recent years, however, it is forests in the tropical and equatorial regions which are dwindling at the fastest rate. In 1990, it was estimated that each year 16 to 20 million hectares of tropical forest have been cleared for agricultural purposes and/or commercial use of timber. The FAO (1986) has estimated that these forests are being removed at the rate of 7.3 million hectares per year, with Brazil contributing the greatest share to tropical deforestation. Current estimates suggest that over 80 per cent of the total biomass burned globally each year is concentrated within the equatorial and subtropical regions (Prinns et al., 1998). Biomass burning associated with deforestation activities is primarily located in developing countries of South America, Africa and South-East Asia, where economic pressures and increased population demands are providing the impetus for increased forest exploitation and resource development.

Biomass burning refers to the burning of the world's forests, grasslands and agricultural lands following the harvest. It is in addition a widespread practice for land clearing and land-use change, and is in particular evidence in the tropical rainforest regions of the equatorial latitudes. Although biomass burning may represent an effective method for clearing land, it has serious environmental consequences on local to global scales. Burning emits large amounts of matter into the atmosphere. Large aerosol particles from burning vegetation spend a short time in the atmosphere but may effectively increase the infrared re-radiation from the lower atmosphere. Small particles remain for periods that may be as long as several weeks or even years in the upper troposphere. The net effect of smoke pollution at the ground surface may be cooling or warming, depending on the direction of changes of the surface albedo and on the absorption coefficient of the particles in the atmosphere. Burning also releases nutrients, especially nitrogen, into the atmosphere and into the soil water. Eventually part of the latter will enter the drainage system and be lost.

The economic, ecological, social and other costs related to deforestation have not been seriously assessed. On the basis of the current rates of deforestation, it is plausible that natural tropical forests will largely disappear by the end of the twenty-first century. However, their conversion into vast expanses of wasteland with little vegetative cover of economic value is not entirely improbable. These changes would imply extensive regional and global changes in climate. Biomass burning contributes significantly to the emissions of atmospheric trace gases such as carbon monoxide, nitrogen oxides and hydrocarbons. For example, current evaluations estimate that about half of the global surface emissions of carbon monoxide result from biomass burning. As these trace species act as ozone precursors, biomass burning plays an important role in the budget of ozone in the troposphere.

The burning of biomass results in the production and release into the atmosphere of significant quantities of gaseous and particulate combustion products. Combustion gases include carbon dioxide, carbon monoxide, hydrocarbons, oxides of nitrogen, methyl chloride and methyl bromide (Granier et al., 1998). Carbon dioxide and methane are important greenhouse gases that affect global climate. Carbon monoxide, hydrocarbons and the oxides of nitrogen are chemically active gases and change the composition and chemistry of the troposphere. Biomass burning is a major global source of elemental carbon particulates, which absorb and scatter incoming solar radiation, and hence alter the Earth's radiation budget and global climate. In addition, biomass burning affects the biogenic production and emission of nitrogen and carbon gases from the soil, and therefore intervenes, in the hydrological cycle.

In addition to regional and global impacts on the climate system, deforestation has a number of other consequences, such as intensified seasonal flooding with resultant loss of lives and property, or water shortages during the dry season. The loss of vegetation cover can lead to the enhanced erosion of agricultural lands and the silting of rivers and coastal waters. The impacts of deforestation on biodiversity are immense.

As in many of the large-scale impacts on mountain regions, deforestation and biomass burning will essentially have indirect effects on mountain environments through climatic change as well as particulate and gaseous pollution transported over long distances. Direct effects of deforestation of mountain trees also occur and are linked to enhanced erosion of denuded slopes and loss of topsoil that could potentially be of agricultural value. Additionally, there is a significant loss of biodiversity in unique mountain forest environments, such as the cloud forests of Central America, Papua New Guinea and East Africa.

The problem of deforestation and biomass burning, and the potential impacts on the global climate system, is a genuinely interdisciplinary one, whose causes often find their origin in economic imbalance between the industrialized countries and the developing world, as well as general environmental mismanagement. The physical mechanisms related to burning impact on the physico-chemical characteristics of the atmosphere, which then feed back into the biospheric systems. Spatial scales over which the effects of biomass burning are important range from the regional to the global scales, over time-frames of a few weeks to a few years. Any mitigation strategies aimed at substantially reducing biomass burning practices will require financial incentives and an international approach to the problem.

5.4.2 Desertification

Climatic changes over a very long time scale can result in the gradual reduction of ground cover, species diversity, humus and mineral content of the soil, and structural characteristics of the vegetation and soil which eventually lead to the formation of deserts. However, human activities often accelerate this process. Overgrazing is the major cause of desertification worldwide. Plants of semi-arid areas are adapted to being eaten by sparsely scattered large grazing mammals which move in response to the patchy rainfall common to these regions. Early human pastoralists living in semi-arid areas replicated this natural system as they moved their small groups of domestic animals in response to food and water availability. Such regular stock movement prevented overgrazing of the fragile plant cover. During the twentieth century, however, the use of fences has prevented domestic and wild animals from moving in response to food availability, and overgrazing has often resulted. The use of boreholes and windmills also allows livestock to stay throughout the year in areas formerly grazed only during the wet season. Where not correctly planned and managed, provision of drinking water contributed to a rapid advance of deserts in the latter part of the twentieth century, as animals gathered around waterholes and overgrazed the area. Incorrect irrigation practices in arid areas can cause salinization that can prevent plant growth. Increasing human population and poverty contribute to desertification as people apply the short-term policy of overexploitation of their environment, without the ability to plan for the long-term effects of their actions.

Once initiated, deserts will tend to spread rapidly. Sand not fixed in position by plants will spread in sandstorms while sand dunes can advance by the action of wind, and encroach upon fertile lands and human settlements. The dynamics of erosion are such that land already eroded by wind and rain becomes highly sensitive to additional erosion. In addition, there are adverse climatic consequences of erosion, because much water which falls out as precipitation is not retained by the soil, but is lost through surface runoff and by enhanced evaporation. Changes in the vegetation cover, ultimately resulting in barren lands, may change a local climate from being mild and relatively moist to one in which the barren surface generates climatic conditions which will exacerbate the conditions leading to aridity. Each year, the desert and arid regions of the world increase by about $200,000\,km^2$, and the process is accelerating. UNEP (1993) has estimated that the direct annual costs resulting from desertification amount to $42.3 billion annually.

Desertification contributes to global climate change by changing the original vegetation cover to an exposed land surface, which has consequences for albedo and therefore the surface energy balance. A barren land surface will have significantly different evaporation and moisture characteristics compared with the original vegetation cover, leading to different exchanges of sensible and latent heat with the atmosphere. As a result of the loss of topsoil by the action of wind, arid regions inject aerosols into the atmosphere. Furthermore, by contributing to a general loss of biomass and plant productivity, desertification disrupts the global biogeochemical cycle and in particular that of the global carbon balance.

5.4.3 Soil degradation

Soil productivity is reduced when there is a loss of soil nutrients, organic matter or moisture content. During the twentieth century, consumption patterns, demography and technology resulted in a large increase in agricultural production, but without a parallel increase in soil conservation measures. This led to a progressive loss of soil fertility and structure, as well as to considerable erosion. To increase agricultural yields, developing countries have had to invest in expensive and energy-intensive techniques which make use of irrigation, mechanization, pesticides and fertilizers. The prevailing system of grants and subsidies which valorize intensive, mechanized farming practices ultimately produce accelerated soil degradation.

The principal causes of soil degradation include inappropriate land use, erosion, salinization and chemical contamination. Reduced protection of the soil by vegetation also makes the area more liable to wind erosion. Soil is usually covered by some form of vegetation that protects the soil from the adverse action of rain or wind. Root systems of plants hold the soil together. Even during periods of drought, the roots of native grasses, which extend several metres into the ground, maintain the cohesion of the soil and keep it from blowing away. When the natural vegetation cover is removed, often through human interference, soil is vulnerable to damage, and the generally slow rate of natural erosion is enhanced. Loss of soil takes place much faster than new soil can be created.

In advanced stages of erosion, all soil, and therefore all capacity for production, may be removed. More frequently, lack of soil conservation results in the loss of the most productive layers of the soil, i.e. those having the highest capacity for the retention of moisture, the highest soil nutrient content, and the most ready response to artificial fertilization.

Most regions which have century-long traditions of agriculture, from the arid Middle East to semi-arid Mediterranean countries, and even in the more temperate mid-latitudes, experience varying degrees of soil erosion. Evidence of erosion is abundant in such culturally diverse areas as southern China, the Indian plateau, South Australia, South Africa, the Russian Federation, Spain, Central America, and the south-eastern and mid-western United States.

In the United States, soil erodes at an average rate of 4-5 tonnes per hectare per year. Between 1972 and 1976 the area cultivated for crops expanded by some 24 per cent, while soil erosion increased far more. By 1976, US farmers were losing an estimated 6 tonnes of soil for every ton of grain they produced. Almost half of America's cropland is losing soil faster than it can be replaced. In developing countries, erosion rates are twice as high as those in the United States, partly because population pressure forces land to be more intensively farmed. Worldwide, farmers are losing an estimated 24 billion tonnes of topsoil each year, while new topsoil production cannot keep up with this loss. The difference between creation and loss represents an annual loss of about 20 tonnes per hectare worldwide, and erosion has now reached unprecedented proportions. In certain regions of the world, where poverty and high population growth rates are common, the loss of soils could severely threaten food security.

These problems may be comparatively marginal for mountains and uplands, but in many developing countries, mountain agriculture represents a substantial source of income on the local to national level. This is particularly the case in Andean and Himalayan countries, as well as in the highland zones of Eastern Africa. Because of the extremely high rates of erosion characteristic of regions of complex orography, the loss of soils through mismanagement and the abandonment of traditional practices has acute consequences for agricultural productivity and food security. Disturbance of centuries-long traditional lifestyles through the loss of a major resource often leads to irreversible damage to the social structure of a region. This is occurring today in even the most remote regions of the world, such as Zanskar in north-west India, where the balance between population growth and resource availability is being destroyed by a progressive shift to non-traditional, non-sustainable agricultural and industrial practices. With instant access to communication technology, and the adverse influence of advertisement and inappropriate technology on traditional agriculture, such regions will be increasingly disrupted in coming decades.

Indirect effects of soil loss for mountains are, as in the case of desertification, the transport of particulate matter (i.e. as a form of pollutant) by atmospheric flows and the deposition of exogenous material in the mountain regions.

5.5 Climatic change

5.5.1 Greenhouse gases

The so-called 'greenhouse gases' (GHGs) are minor gaseous constituents which have radiation properties capable of warming the atmosphere. A fraction of the solar energy which is intercepted at the top of the atmosphere reaches the ground and warms the surface. The extent of direct solar heating depends on a number of factors, such as the reflectivity of the surface. In order to avoid continuously absorbing energy and overheating, the Earth emits this energy to space in the form of infrared radiation. GHGs absorb infrared radiation in certain wavebands of the infrared electromagnetic spectrum, and re-emit this heat energy in all directions, including the atmosphere and the Earth's surface. In doing so, GHGs maintain low-level atmospheric temperatures at a level 33°C higher than would otherwise be the case. In the absence of trace gases such as carbon dioxide, the temperature of the Earth would be about −18°C on average. Greenhouse gases are therefore life-sustaining. They represent less than 5 per cent of the gaseous composition of the atmosphere; that is, the gases which are climatically relevant make up a very modest proportion of the atmosphere.

Human activity, through industry, agriculture, energy generation and transportation, has released significant amounts of GHGs into the atmosphere since the beginning of the industrial era, and there is concern that these gases may be modifying the global climate through an enhancement of the natural greenhouse effect. According to the IPCC (1996), global mean temperatures could increase by 1.5 to 4.5°C by the end of the twenty-first century in response to this additional radiative forcing. While this may appear to be a minor warming when compared to diurnal or seasonal amplitudes of the temperature cycle, it should be emphasized that this is a warming unprecedented in the past 10,000 years. It is not only the amplitude of change but also the rate of warming which is generating concern in the scientific community, especially in terms of the vulnerability and response of environmental and socio-economic systems to climatic change.

The largest contributor to the natural greenhouse effect is water vapour. Its presence in the atmosphere is not directly affected by human activity, although it is likely to increase as a result of global warming and therefore enhance the greenhouse effect ('positive feedback'). This is because warmer air can hold more moisture; however, the complex interactions between clouds, radiative processes and climate are still not fully understood, and thus the precise magnitude of this feedback remains unknown. Carbon dioxide is currently responsible for over 55 per cent of the anthropogenic greenhouse effect, as illustrated in Figure 5.5. All processes which involve combustion of fossil fuels (industrial activity, transportation and energy use) release carbon into the system; in addition, deforestation releases carbon stored in trees. Current annual emissions amount to over 7 Gt of carbon, which represents an annual increase of about 1 per cent in CO_2 concentrations. Anthropogenic emissions of CO_2 enter the natural carbon cycle, and large quantities of carbon are exchanged naturally each year between the atmosphere, the oceans and terrestrial vegetation. In the absence of human interference, the carbon in the system was in approximate balance; CO_2 levels have fluctuated within a

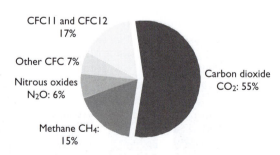

FIGURE 5.5 Principal anthropogenic greenhouse gases.

10 per cent margin since the end of the last glaciation. Since the beginning of the industrial era in the early 1800s, however, levels have risen by almost 30 per cent. The observed increase in CO_2 concentrations since this period is an indication that human activities have disturbed this equilibrium. Only about 50 per cent of anthropogenic carbon dioxide emissions are absorbed by the oceans and the biosphere.

Methane (CH_4) is a powerful greenhouse gas whose levels have doubled since the beginning of the twentieth century. The principal sources of methane are agricultural, notably in the tropical regions (rice), livestock (particularly cattle), and emissions from other concentrated organic material such as in waste dumps. Methane currently contributes 15–20 per cent of the greenhouse effect. Nitrous oxide (N_2O), chlorofluorocarbons (CFCs) and ozone contribute the remaining 25–30 per cent of the enhanced greenhouse effect. N_2O is an agricultural by-product, mainly formed in the digestive tract of livestock and from fertilizers. CFCs are entirely artificial gases that are found in refrigerants, solvents and spray propellants. Because of their responsibility in the chemical destruction of stratospheric ozone, CFC levels have stabilized in the 1990s as a result of stringent emission controls introduced under the 1987 Montreal Protocol.

Aerosols are microscopic particles (generally sulphur dioxide, SO_2) emitted in certain types of industrial processes, by coal-burning power stations, and also through biomass burning. Aerosols represent an additional anthropogenic forcing factor, but are not a GHG. They are removed from the atmosphere by gravity and precipitation scavenging after only a few days, but they are emitted in such large quantities that they have a substantial impact on climate. They locally cool the atmosphere and the surface by scattering sunlight back to space. Over heavily industrialized regions, and those which are located downwind of the dominant atmospheric flows, aerosol cooling is capable of counteracting much of the warming effect of GHGs. Aerosols do not have a global effect, as does CO_2, for example, because they are not evenly mixed into the atmosphere, as are the gaseous GHGs.

The warming potential is different from one GHG to another, molecule for molecule. CO_2 is by no means the most efficient GHG in terms of warming potential and CH_4, for example, has a warming potential roughly 60 times greater, N_2O of the order of 250 times, and certain CFCs over 6000 times. At present, atmospheric concentrations of all gases other than CO_2 are still very low. However, significant increases in methane levels in the future, for example, would result in a larger share for this gas in terms of atmospheric warming.

A further problem related to GHGs is their relatively long residence times in the atmosphere, as shown in Figure 5.6. Gases such as CO_2 have residence times of the order of 1–2 centuries. Methane is relatively short-lived, of the order of a decade or so, while the CFCs have century-scale lifetimes. This is a result of either the very strong chemical bonds of these molecules, which do not disintegrate rapidly, or low recycling rates in which emissions

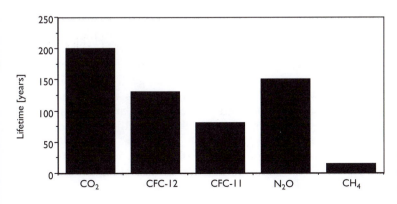

FIGURE 5.6 Typical residence times of the major greenhouse gases

TABLE 5.3 IPCC scenarios for atmospheric CO_2 concentrations by the year 2100 (in ppmv)

IS92A	IS92B	IS92C	IS92D	IS92E	IS92F
733	710	485	568	986	848

Note: The 1990 figure was 355 ppmv (parts per million by volume)

exceed absorption. Figure 5.6 implies that even if all GHG emissions are halted in the near future, their influence will continue to be felt for a century or more.

The future evolution of atmospheric GHG concentrations resulting from human activity is dependent on population growth, economic activity, and energy and industrial consumption, as noted earlier in this chapter. The IPCC has provided estimates of the possible trends in GHG concentrations to the end of the twenty-first century, as shown in Table 5.3 for the six basic scenarios IS92A–F.

Low population growth and low economic growth lead to a relatively modest rise in GHG levels by the end of the next century, which nonetheless reaches a doubling of pre-industrial CO_2 levels. The 'worst-case' scenario leads to CO_2 concentrations close to 1000 ppmv, which is almost a fourfold increase over pre-industrial levels. It is likely that we are committed to at least a doubling of GHG concentrations sometime during the next century; stringent reductions in emissions and the creation of new sinks of GHGs, for example through afforestation, would be required in order not to attain levels higher than 500 ppmv.

Linked to climatic change is the problem of stratospheric ozone depletion; at altitudes of between 25 and 45 km in the atmosphere, ozone filters incoming ultraviolet (UV) radiation before it reaches the surface. Exposure to UV radiation can be damaging to life on Earth, and a small decrease in ozone concentrations can lead to higher incidence of UV-induced hazards, such as skin cancers and cataracts in humans, and irreversible damage to the cellular structure of plants. Ozone depletion became widely reported in the late 1980s with observations of the yearly formation of

the 'ozone hole' over the Antarctic. More recently, European and American scientists have reported significant decreases in ozone concentrations over northern latitudes, and NASA satellite photographs have shown that ozone losses in mid-latitudes, including the Alpine regions, can periodically reach 10 per cent. The CFCs, which are used in a variety of industrial applications such as refrigerants, solvents and propellants, are responsible for stratospheric ozone depletion through complex chemical reactions. The link to global climate change is that the CFCs are also greenhouse gases with long residence times, as shown in Figure 5.6, and their atmospheric warming potential far exceeds that of other GHGs such as carbon dioxide.

5.5.2 The greenhouse forcing of climate

General circulation model (GCM) simulations of the response of global climate to enhanced GHG concentrations have been a key focus of the IPCC reports (IPCC, 1996). All model simulations point to a warming which on a global average is in the range of 1.5–4.5°C in 2100, according to the scenario used. The sensitivity of climate to a doubling of CO_2 yields a temperature increase of about 2.5°C compared to current climate.

Many climate modelling groups make use of coupled ocean–atmosphere simulations, in which the ocean and atmospheric components of the system dynamically interact. Typical results for one such model, the MPI ECHAM4 GCM, are illustrated in Figure 5.7. The global mean temperature increase in this experiment with respect to the control simulation is 1.5°C (Beniston, 1997). It is seen that the warming is largest over the continents and at high latitudes. The land surface is responding faster to the greenhouse forcing than the ocean owing to its large thermal inertia. At high latitudes, reductions in seasonal snow cover and sea-ice extent and the associated positive feedback lead to the amplification of the greenhouse forcing at higher latitudes. The increase in the Arctic is strongest during autumn; this can be explained by a reduced season for sea-ice expansion.

Box 5.1 Greenhouse gas (GHG) emission scenarios

Most scenarios suggest that future growth in emission rates will be dominated by social and economic evolution in developing countries. The bulk of emissions to date have come from industrialized countries. However, most future growth is likely to come from emerging economies where economic and population growth is fastest – and for which projections are most uncertain.

In a typical scenario in which no restrictive policies are adopted to reduce emissions (IS92A), CO_2 emissions will rise from 7 Gt of carbon annually in 1990 to 20 Gt in 2100. World population doubles by 2100 while economic growth continues at 2–3 per cent per year.

This scenario leads to the equivalent of a doubling of pre-industrial CO_2 concentrations by 2030, and a trebling by 2100. This includes the effects of other greenhouse gas emissions, translated into their carbon dioxide equivalents. Even a doubling of pre-industrial carbon dioxide would represent levels of long-lived greenhouse gases higher than they have been for thousands of years.

Different assumptions about sources and sinks give very different results. Future emissions are uncertain, and they have to be translated into future atmospheric concentrations using models of the carbon cycle and atmospheric chemistry. This introduces more uncertainty, since it is unclear how sinks such as the oceans and the biosphere may respond to a changing climate. Rising CO_2 levels, for example, increase the primary productivity of plants, in what

is often referred to as the CO_2 fertilization effect, and absorb more carbon dioxide through photosynthesis.

Intervention scenarios are designed to examine the impact of policies implemented at an international level in order to reduce GHG emissions. Scenarios IS92C and D, for example, depend on assumptions not only about population and economic growth, but also about how future societies will respond to the introduction of policies such as taxes on carbon-rich fossil fuels. Existing international commitments could reduce the rate of growth in emissions in coming decades. The commitment made at the Kyoto Conference of the Parties to the United Nations Framework Convention on Climate Change (UN-FCCC) in December 1997 to reduce GHG emissions over the next decade by an average 5.6 per cent is one small step in this direction. Stabilizing atmospheric GHG concentrations would require all countries to make very substantial cuts in their emissions, to take them well below current levels. For example, stabilizing carbon dioxide at double its pre-industrial concentration sometime in the twenty-second century would require emissions to fall eventually to less than 30 per cent of their current levels, despite growing populations and an expanding world economy.

It is necessary to stress here that stabilizing emissions is not equivalent to stabilizing concentrations. Stabilizing emissions simply means that rates of emission of GHGs will not continue to rise; GHGs will continue to be injected into the atmosphere, and unless the emissions are in approximate balance with the sinks of GHGs, then concentrations will continue to rise.

Figure 5.8 shows the shifts in precipitation that this particular GCM simulates. Certain regions in the tropics experience an increase in precipitation as a result of an enhancement of the hydrological cycle that is physically consistent with a warmer climate. This is

particularly the case over the Amazon basin, and in the monsoon-prone Indian subcontinent. Other regions, particularly in the mid-latitudes, exhibit a conspicuous decrease as storm systems become less pronounced in the future as a result of a reduced temperature

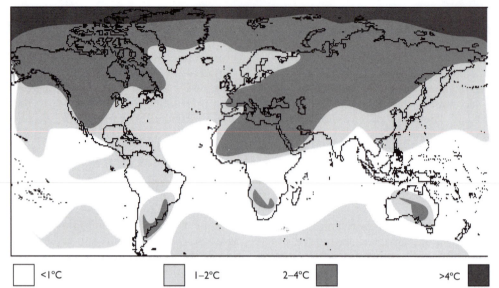

☐ <1°C   ▨ 1–2°C   2–4°C ▨   >4°C ■

FIGURE 5.7 General circulation model (GCM) scenario of global warming for doubled CO_2 concentrations compared to pre-industrial levels

gradient between the poles and the equator. The reduction in precipitation associated with frontal systems is particularly marked over the North Atlantic and into North-Eastern Europe, as well as over the North Pacific.

Both temperature and precipitation changes, as illustrated in Figures 5.7 and 5.8, will have consequences for mountain hydrological, cryospheric, biospheric and socio-economic systems. The scale of the maps is

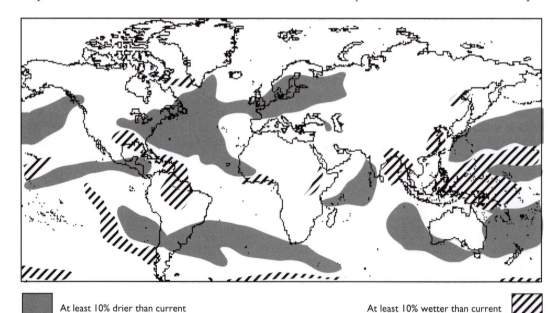

▨ At least 10% drier than current   At least 10% wetter than current ▨

FIGURE 5.8 General circulation model (GCM) scenario of precipitation change for doubled CO_2 concentrations compared with pre-industrial levels

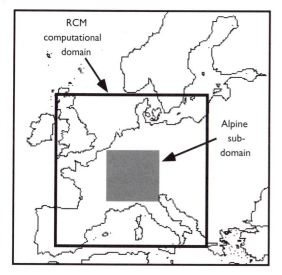

FIGURE 5.9 Regional climate model (RCM) domain applied to climatic change in the Alps

such, however, that it would be difficult to enter into any significant level of detail concerning changes at the regional scales. It is for this reason that techniques are required to provide information on the response of regional environmental systems to changes in global climate forcing, as discussed in Chapter 4.

As an example of regionalization of global model results, the nested modelling techniques (*see* Section 4.3.3), whereby GCM simulations are used as boundary and initial conditions to an RCM (regional climate model), are applied to the Alpine region of Europe. These simulations, reported by Beniston *et al.* (1995), Marinucci *et al.* (1995) and Rotach *et al.* (1996), make use of the MPI ECHAM4 GCM as the driving global model developed at the Max-Planck Institute for Meteorology, Hamburg, Germany, and the RegCM2 RCM developed at the National Center for Atmospheric Research, Boulder, Colorado. The RCM operated with a 20-km resolution over Western Europe, centred over the Alps, with a total grid domain of 1100 km × 1100 km. Figure 5.9 illustrates the geographic distribution of the RCM domain over Europe.

The 20-km grid does not adequately perceive the fine detail of Alpine orography, but nevertheless allows an improved resolution

of the presence of the Alps and the major valley and mountain systems, compared to the GCM grid. Using the ECHAM4 results for the double-CO_2 'time window', it was possible to establish statistics, based on the physics of the system, concerning the salient features of the regional response of Alpine climate to global GHG forcings. Typical results, in the form of difference maps between future climate under enhanced atmospheric GHG concentrations and current climatic conditions, are illustrated in Figure 5.10, for both winter (January) and summer (July) conditions. It is seen that for both seasons, temperature increases compared to current climate, by about 1.5°C in winter and up to 4°C or more in summer. Precipitation conditions, according to these simulations, however, differ between winter (where there is an increase) and summer. The sharp reduction in precipitation in summer explains the much stronger warming than in winter: cloudiness is reduced, and therefore more incoming solar energy is available to warm the surface; in addition, soil moisture diminishes, thereby exerting a positive feedback on the lower atmosphere.

These results need to be viewed with caution, however, because of the large uncertainties involved in the nested modelling approach. There are problems of spatial representativeness, even at the 20-km grid scale used here; the grid does not capture the finer details of the Alpine topography, which are important in determining regional climate characteristics. Other problems include the propagation of GCM errors into the RCM domain, such as a potentially too large drying out of the Mediterranean region in summer as simulated by the ECHAM4 GCM. These overly dry climatic conditions have repercussions for the manner in which the Alpine zone responds, since a signal of exceptional drying and warming for the Mediterranean zone as projected by the GCM drives the RCM simulations for the summer period. If, for numerous reasons, the GCM results turn out to be in error, these errors will be reflected in the RCM projections for the Alps. It should be noted, however, that anomalies of 4–5°C in the future for a particular month are not

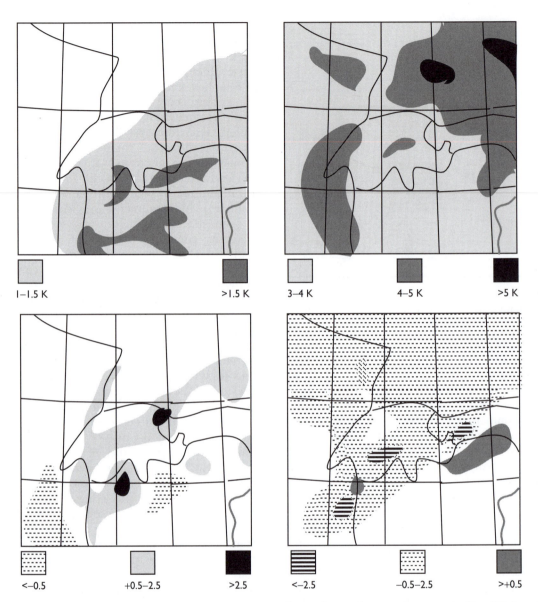

Figure 5.10 Differences in precipitation between an enhanced greenhouse gas scenario ($2 \times CO_2$) and current climate, as simulated by a regional climate model (RCM) centred over the European Alps. Upper left: ΔT for January (°C); upper right: ΔT for July; lower left: precipitation change in January (mm); lower right: precipitation change in July

necessarily absurd. Recent statistics show that some months exhibit large departures from mean conditions, e.g. December 1997. Temperatures for this month in certain parts of the Alps were over 8°C warmer than the climatological average (SMA, 1997).

Impacts of environmental change on natural systems

6.0 Chapter summary

The potential impacts of changes in environmental conditions on natural mountain environments are discussed here. The rationale and the challenges for impact assessments are first addressed, before the manner in which hydrological and cryospheric systems may change in the twenty-first century and beyond is examined in closer detail. The response of various mountain ecosystems is also considered, in particular that of forests and plants, soils and other forms of disturbance such as fire. It is seen that many of the interactions between the forcing factors responsible for change, such as climate, and the impacted systems are exceedingly complex. A single causal mechanism such as climatic change is unlikely to be the only stress factor, and will tend to exacerbate the stresses exerted by other environmental conditions and direct human interference. Because of the inherent fragility of mountain environments, it is almost impossible for systems which have been significantly disturbed to revert to their initial state.

6.1 Challenges for impact assessments

Questions which need to be posed in the context of impact assessment studies include the sensitivity of a given system to a change in its environment, the vulnerability of that system to change, and its capacity for adaptation. *Sensitivity* of a system is an issue that requires adequate knowledge of its current conditions for sustainability, and the range of stresses that it can withstand without significant damage. *Vulnerability* of a system is the extent to which it may be damaged by changing environmental stresses; this may include not only the magnitude but also the rate of change. *Adaptability* of a system describes the degree to which ecological and socio-economic systems may adjust to changes in their environment; in many cases, natural and social systems have the capability to resist the adverse consequences of change or to benefit from new opportunities which environmental change may provide.

There are a number of challenges when dealing with such issues, which are all related to the dominantly regional or local characteristics of the systems upon which environmental change is likely to have an influence. In terms of climatic change, there is limited confidence in regional model projections of relevance to impacts; uncertainty in model results is known to increase as numerical model resolution is refined. There is the added complexity due to the fact that patterns of anthropogenic climatic change are superimposed on natural climate variability. Furthermore, regional climate response to global climate forcing will be influenced not only by greenhouse gases but also by aerosols, which have only recently begun to be simulated by general circulation models (GCMs) (IPCC, 1996).

When analysing the potential effects of change on a particular system, it should be borne in mind that environmental change comprises a range of stresses. For example, the IPCC (1996) has emphasized in its Second Assessment Report that human-induced climatic change represents an *additional stress* on systems that are already threatened by *other changes*. These include diminishing biodiversity, overexploitation of natural resources, numerous forms of air, water and soil pollution, as well as ozone depletion in the stratosphere. Because of this complex interlinkage of stresses, it is exceedingly difficult to attribute impacts on natural or economic systems to any one particular element of environmental change, e.g. climatic change. As a result, the development of adaptation or mitigation strategies in the face of the uncertainties involved is a daunting task for the policy-making sphere.

An additional challenge which impacts research poses to policy-makers is related to the interpretation and implementation of specific regulations of conventions dealing with aspects of environmental change. In its Article 2, the UN Framework Convention on Climate Change (FCCC), for example, stipulates that stabilization of atmospheric greenhouse gas concentrations should remain below a threshold

> that would prevent dangerous anthropogenic interference with the climate system. Such a level should be achieved within a time-frame sufficient to allow ecosystems to adapt naturally to climate change, to ensure that food production is not threatened and to enable economic development to proceed in a sustainable manner.

The challenges presented to the policy-maker by Article 2 are the assessment of the concentrations of greenhouse gases which can be regarded as *dangerous*, what constitutes *natural adaptability* of ecosystems (some will be very sensitive to small shifts in climatic conditions while others may hardly respond to large changes), and how to shift from current economic development patterns to those which are based on *sustainability*.

Few assessments of the impacts of environmental change in general, and climatic change in particular, have been conducted in mountain regions, as opposed to other biomes such as tropical rainforests, coastal zones or high-latitude or arid areas. This is mainly because, as already mentioned, mountain orography is often too poorly represented in GCMs and even in RCMs for meaningful projections to be applied to impacts models. There is also a significant lack of comprehensive multidisciplinary data for impact studies, which are among the prerequisites for case studies of impacts on natural or socio-economic systems (Kates *et al.*, 1985; Riebsame, 1989; Barry, 1992b; Parry *et al.*, 1992). In addition, the complexity of physical, ecological and social systems in mountains, and their mutual interdependency, pose significant problems for assessment, particularly because there is often little or no quantification of the value of mountains in monetary terms, as Price (1990) has pointed out.

Despite these severe constraints, it is possible to assess to a certain extent the manner in which environmental change may impact various biogeophysical systems, on the basis either of RCM scenarios such as those discussed in Chapter 5, or of a 'what if' approach, which determines the sensitivity and vulnerability of a system to a hypothetical change or range of plausible changes.

6.2 Impacts on hydrology

Hydrological systems are controlled by soil moisture, which also largely determines the distribution of ecosystems, groundwater recharge and runoff; the latter two factors sustain river flow and can lead to floods. These controls are themselves governed by climate, and hence any shifts in temperature and precipitation will have significant impacts on water. Increased temperatures will lead to higher rates of evaporation, and a greater proportion of liquid precipitation compared to solid precipitation; these physical mechanisms, associated with potential changes in precipitation amount and seasonality, will affect soil moisture, groundwater reserves, and the frequency of flood or drought episodes. Though water is present in ample quantity at the Earth's surface, the supply of water is limited and governed by the renewal processes associated with the global hydrological cycle. Mountains can intercept substantial amounts of water in the form of snow and rain from the predominant large-scale atmospheric flows. This water ultimately flows out as rivers, forming the most accessible supply of fresh water to the more densely populated plains downstream of the mountains. With the expansion of human settlements and the growth of industrial activities, water has been increasingly used for the assimilation and discharge of wastes. This resource has been taken for granted, and only in the past few decades have increasing water shortage and declining water quality (because of pollution) drawn attention to the inherent fragility and scarcity of water. This has led to concerns about water availability to meet the requirements of the new century.

According to Shiklomanov (1993), the global annual water demand is likely to increase from $4130\,km^3$ in 1990 to $5190\,km^3$ in 2020, if present consumption patterns are sustained. Because of increasing population, the additional demand will be accompanied by a sharp decline in water availability per capita. While a consumption of $1000\,m^3$ of water per year and per capita is considered a standard for 'well-being' in the industrialized world, projections of annual water availability per capita by the early twenty-first century for North Africa are $210\,m^3$, for Central Asia and Kazakhstan $700\,m^3$, and for South Asia $1100\,m^3$. This trend is declining in all parts of the world, including those that are considered to have ample water resources. This in turn has prompted the need for a more rational approach to the conservation and use of what is probably the most vital single resource for humankind. An additional source of concern is that mountains have long been considered as an exclusive source of water for the lowland populations. New initiatives are aimed at conservation and distribution of water within the mountains of the developing world so that mountain people, in particular women, can avoid spending a large part of their working lives merely carrying drinking water for their families.

Against this backdrop of social problems, it is obvious that water resources will come under increasing pressure in a changing global environment. Significant changes in climatic conditions will affect demand, supply and water quality. In regions which are currently sensitive to water stress (arid and semi-arid mountain regions), any shortfalls in water supply will enhance competition for water use for a wide range of economic, social and environmental applications. In the future, such competition will be sharpened as a result of larger populations, which will lead to heightened demand for irrigation and perhaps also industrialization, at the expense of drinking water (Noble and Gitay, 1998). It would be hazardous to assume that present-day water supply and consumption patterns will continue in the face of increasing population pressures, water pollution, land degradation and climatic change. Events in recent history may provide useful guidelines for developing such strategies (Glantz, 1988).

Projections of changes in precipitation patterns are tenuous in GCMs, even those operating at high spatial resolution, because rainfall or snowfall are difficult variables to simulate, compared to temperature. A number of assessments of the potential impacts of climate change on water resources, including snowfall and storage, have been conducted at a variety of spatial scales for most mountain regions, as reported by Oerlemans (1989), Rupke and Boer (1989), Lins et al. (1990), Slaymaker (1990), Street and Melnikov (1990), Nash and Gleick (1991), Aguado et al. (1992), Bultot et al. (1992) and Leavesley (1994). Riebsame et al. (1995) have found that in many cases, it is difficult to find changes in annual river flows in response to climatic change, but that seasonality changes were often detected.

It has been recognized more recently that the superimposed effects of natural modes of climatic variability (e.g. El Niño/Southern Oscillation – ENSO – or the North Atlantic

Oscillation), which can perturb mean precipitation patterns on time scales ranging from seasons to decades, are important mechanisms to take into account but are not well predicted by GCMs. The warm or cold phases of ENSO (respectively, El Niño and La Niña) tend to reverse normal precipitation patterns, particularly in the countries bordering the Pacific, as well as in other parts of the world, such as Eastern and Southern Africa. Higher than average precipitation has strong impacts on stream flow in river basins which are not accustomed to excessive rainfall; this has often been seen to be the case in the arid

regions of the Andes and coastal ranges in northern Chile and Peru, where catastrophic floods frequently occur during such events. In Costa Rica, some of the more severe ENSO events of recent decades have resulted in reduced surface runoff and a higher demand for thermal energy production (Campos *et al.*, 1996).

In Africa, where over 60 per cent of the continent's rural population and 25 per cent of its urban population do not have access to safe drinking water (Zinyowera *et al.*, 1998), climatic change will exacerbate the current situation. The expected degradation will in

---

Box 6.1 Water resources in Turkey, Syria and Iraq: a situation for conflict

Turkey is well endowed with water resources; its mountains and high plateaux intercept much of the available rainfall in this region of the Middle East. Turkey controls the headwaters of the Tigris and Euphrates rivers, which flow through Syria and Iraq. In order to augment its agricultural capacity, both for its increasing domestic needs and for exportation, a large-scale development programme called GAP (the Turkish acronym for 'Big Antatolian Project') is now under way. The project involves converting the current steppes, which are capable of sustaining only a limited number of commercial species such as olive trees and pistachio nuts, to a vast orchard and vegetable-growing area. By 2015, it is planned to construct 21 dams and hydropower stations; the dams would be capable of providing irrigation water for about 1,000,000 hectares drawn from the Euphrates and 600,000 hectares from the Tigris. For both the downstream neighbours Syria and Iraq, the GAP programme is a serious threat to their water supply. The Atatürk dam on the Euphrates has reduced the flow of the Euphrates downstream by 30 per cent to about $300 \, m^3/s$; Syria's own requirements for freshwater supply and

irrigation are estimated at $700 \, m^3/s$, significantly more than what Turkey was delivering even before the completion of the Atatürk dam ($500 \, m^3/s$). Such a situation has a serious potential for conflict (Gleick, 1994), because neither Syria nor Iraq has the means of taking retaliatory measures, e.g., by depleting groundwater reserves. Syria could theoretically engage in an armed conflict with Turkey, but the latter country has a much larger military potential and is a member of the North Atlantic Treaty Organization (NATO), and is thus a strategic ally for the United States. Furthermore, a military cooperation agreement between Turkey and Israel was signed in 1997, with the containment of Syria a major objective. Rather than engage in direct confrontation, Syria could back secessionist movements in the Kurd minorities who are openly hostile to both the Turkish and the Iraqi governments; Syria has in the past provided support to the PKK, a leftwing Kurdish liberation movement, which was suspended in part to try to obtain more water from Turkey. In the absence of international agreements for the sharing of a vital resource in this climatically marginal region, tensions will undoubtedly rise in coming years in reaction to Turkey's dominant position at the headwaters of some of the major rivers of the region.

part be due to increasing population pressures and insufficient financial resources to ameliorate the existing problems of water supply and quality. Furthermore, Africa has a large number of rivers that cross or form international boundaries, so that the sharing of a dwindling resource could ultimately lead to regional conflicts. Armed disputes over water resources may well be a significant social consequence in an environment degraded by pollution and stressed by climatic change, as in the Middle East (Turkey and Syria, as discussed in Box 6.1, or Israel and its Palestinian neighbours).

Impact studies for specific river basins of the world have highlighted the fact that water availability will decline. Sharma *et al.* (1996) have shown that the number of countries which currently suffer from water stress and amount will more than double by 2025, on the basis of the IPCC (1992) climate scenarios. The World Bank (1995) estimates that about 600 million people in Africa will suffer from water shortage and dwindling water quality. Figure 6.1 illustrates the probable reductions in water availability for a number of the countries discussed in the context of Table 1.1. In some of the examples shown in this figure, the water availability per capita decreases below the $1000\,m^3$ threshold discussed earlier. Shifting precipitation belts account for only a fraction of this reduction; rapid population growth, urbanization and economic expansion place additional burdens on water supply. The impacts of water shortages on economic activity can be assessed on the basis of recent droughts; these include agricultural shortfalls in South Africa and Zimbabwe, and power rationing and associated disruption of Kenyan manufacturing and engineering industries (UNEP, 1997).

In Latin America, which accounts for 35 per cent of global non-cryosphere fresh water, the impacts of climate change are expected to occur in the more arid regions of the continent, which are often associated with the rain-shadow influences of the Andes ranges. Shifts in water demands will depend on population growth, industrial expansion and agricultural potential. In many countries of the region, water availability is expected to decline, which is likely to generate potential for international conflicts. The IPCC (1998) estimates that water availability per capita and per annum will decrease from $4750\,m^3$ in 1990 to $2100\,m^3$ in Mexico by 2025 without any change in climate, i.e. due to population growth and economic growth. Based on several GCM simulations, projected shifts in precipitation in a warmer climate yield a range of $1740-2010\,m^3$. For Peru, the respective set of figures are $1860\,m^3$, $880\,m^3$ resulting from demography alone, and $690-1020\,m^3$ with climatic change, i.e. close to or below the minimum requirements for 'well-being'.

Water resources in tropical Asia are very sensitive to tropical cyclones and fluctuations in their trajectories and intensity. The dominant effect of the monsoon may be perturbed in a changing climate. Runoff in the Ganges, for example, is more than six times that of the dry season. As elsewhere in the world, water

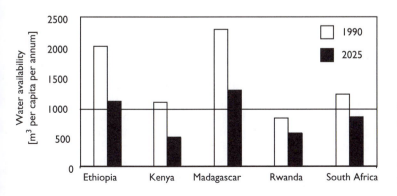

FIGURE 6.1 Water availability in selected African mountain countries in 1990 and (projected) 2025. The $1000\,m^3$ threshold is indicated to show the worsening situation in many countries

resources will become increasingly vulnerable to increasing population growth, urbanization, industrial development and agriculture, as shown by Schreir and Shah (1996). An impact assessment study by Mirza (1997) for a number of Himalayan basins contributing to the Ganges has shown that changes in the mean runoff in different sub-basins ranged from 27 to 116 per cent in a climate forced by a doubling of CO_2 concentrations relative to their pre-industrial levels. The sensitivity of basins to climate change was seen to be greater in the drier catchments than in the wetter ones. However, water demand is greatest in the dry season in India, and demand cannot be met by supply in this season. Shifts in the timing and intensity of the monsoon, and the manner in which the Himalayan range intercepts the available precipitable water content of the atmosphere, will have major impacts on the timing and amount of runoff in river basins such as those of the Ganges, the Brahmaputra or the Irrawaddy. Divya and Mehrotra (1995) and Mirza and Dixit (1997) have shown that amplification or weakening of the monsoon circulations would indirectly impact upon agriculture and fisheries, freshwater supply, storage capacity and salinity control.

Green *et al.* (1997) have discussed indirect influences of the impacts of changing climate on hydrological systems, through associated changes in vegetation. For example, groundwater recharge can be damped or amplified by the dynamic response of vegetation, which in turn contributes to determining the amount of flow which will take place in river basins. Deforestation and urbanization, by changing the physical characteristics of mountain slopes, will tend to increase the amount and speed of runoff, thereby enhancing the risk of flooding, high sediment loads and water pollution. As shown by Bruijnzeel and Critchley (1994) and Campos *et al.* (1996), deforestation in mountain watersheds enhances the sediment load of rivers through increased slope erosion. Higher sediment loading in rivers has negative consequences for water supply and use, particularly when these are regulated by dams, which will silt up far more rapidly than predicted by their design lives.

6.3 Impacts on the mountain cryosphere

Changes in the mountain cryosphere will have a number of indirect consequences; in terms of water supply, changes in seasonal snow-pack and glacier melt will influence discharge rates and timing in rivers which originate in mountains. In terms of tourism, the negative impacts of lack of snow in winter, and the perception of landscape changes in the absence of glaciers and snow, may deter tourists from visiting certain mountain regions.

In most temperate mountain regions, the snow-pack is close to its melting point, so that it is very sensitive to changes in temperature. As warming progresses in the future, current regions of snow precipitation will increasingly experience precipitation in the form of rain. For every 1°C increase in temperature, the snow line rises by about 150 m; as a result, less snow will accumulate at low elevations than today, while there could be greater snow accumulation above the freezing level because of increased precipitation in some regions. Shifts in snow-pack duration and amount as a consequence of sustained changes in climate will be crucial to water availability for hydrological basins, as Steinhauser (1970) has shown.

Some attempts at modelling the snow-pack have been undertaken to determine how this may change in a warmer global climate. Martin and Durand (1998) have used a snow model to investigate the manner in which snow amount and duration are likely to alter in the French Alps under changing climatic conditions. Using the SAFRAN-CROCUS snow model, incorporated into the French ARPEGE GCM, an assessment of the most sensitive areas of the French Alps has been undertaken. Figure 6.2 provides an overview of the reduction in snow amount in the selected regions of the Alps for a particular warming scenario. Sensitivity studies show that lower elevations, i.e. below 1500 m, will respond

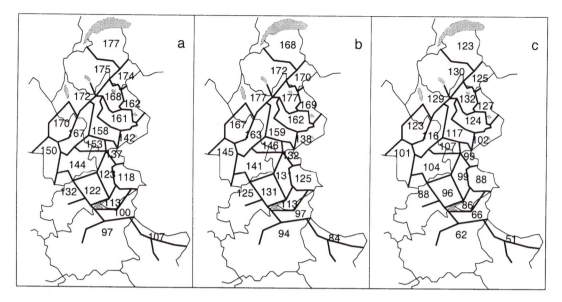

FIGURE 6.2 Changes in snow amount in the French Alps for a climatic change scenario
Source: Martin and Durand (1998)

rapidly to small changes in temperature, especially in the southern part of the French Alps.
Changes in precipitation amount influences
the maximum snow depth more than snow-
cover duration (Martin, 1995). Reduced snow
cover will have a number of implications,
including early seasonal runoff, lower soil
moisture and drier vegetation in summer,
which is likely to lead to greater fire hazard.
Martin and Durand (1998) note that there are
large uncertainties concerning high precipitation, which are linked to the quality of GCM
results. Biases exist at the regional scale for
temperature, and the zonality of the atmospheric circulation, if erroneous, will lead to
precipitation events outside the areas which
are most affected. The variability of the atmospheric circulation is also a major factor which is
underestimated by GCMs; such variability
results in lee cyclogenesis in the western Mediterranean, which is associated with heavy
snowfall events in the southern Alps. Another
important point to be taken into account, if the
model procedure is to reproduce correctly the
observed snow-pack and to simulate its
future evolution, is the decadal-scale variability of regional climate in the Alps.

Glaciers are possibly the most sensitive
system to climatic change, because any
changes in the ratio of accumulation to ablation
of snow and ice, which are dependent on temperature and precipitation, will trigger glacier
mass wasting. Glacier behaviour thus provides
some of the clearest signals of ongoing warming trends related to the enhanced greenhouse
effect (Haeberli, 1990; Wood, 1990; WGMS,
1993). Haeberli (1994) suggests that current
glacier retreat is now beyond the range of
natural variability as recorded during the
Holocene.

The effects of temperature and precipitation
changes on glaciers are complex and vary by
location. In polar latitudes and at very high
altitudes in mid-latitudes, atmospheric warming does not directly lead to mass loss through
melting/runoff but to ice warming (Robin,
1983). In areas of temperate ice which predomi-
nate at lower latitudes or altitudes, atmospheric warming can directly impact the mass
and geometry of glaciers (Haeberli, 1994).
Empirical and energy-balance models indicate
that 30–50 per cent of existing mountain glacier
mass could disappear by the 2100 if global
warming scenarios indeed occur (Oerlemans

and Fortuin, 1992; Kuhn, 1993; Haeberli, 1995; Fitzharris *et al.* 1996; Haeberli and Beniston, 1998). The smaller the glacier, the faster it will respond to changes in climate. With an upward shift of 200–300 m in the equilibrium line between net ablation and net accumulation, the reduction in ice thickness of temperate glaciers could reach 1–2 m per year. As a result, many glaciers in temperate mountain regions would lose most of their mass within decades (Maisch, 1992).

Impact studies for different mountain regions of the globe confirm these projections. In East Africa, for example, Hastenrath and Rostom (1990) have shown the Lewis and Gregory glaciers on Mt Kenya to be in constant regression since the late nineteenth century. Should this trend continue as expected in the twenty-first century, there would be very little permanent ice left in this region within the next 100 years. Similarly, Schubert (1992) has used photographic evidence from the nineteenth century to show that the snow line in Venezuela has risen from 4100 m to more than 4700 m today. These changes in ice extent and in the snow line altitude have had important geo-ecological effects, leading to shifts in vegetation belts and to the fragmentation of previously continuous forest formations. Enhancement of the warming signal in these regions would lead to the disappearance of significant snow and ice surfaces. In north-west China, projected changes in precipitation and temperature over the next century are likely to lead to the disappearance of one-fifth of glacier surfaces in this region (Wang, 1993).

A consequence of increased glacier mass ablation in coming decades is that there will be enhanced runoff as the ice disappears; this extra runoff could persist for a few decades to a few centuries, according to the size of the melting glaciers. Use could be made of this resource in terms of hydro-power or irrigation in the interim period until the glaciers disappear completely. Yoshino *et al.* (1988), for example, indicate that runoff from glaciers in Asia could triple in volume by the middle of the twenty-first century in response to global warming as projected by the IPCC (1996).

Permafrost will also respond to climatic change, although investigations of mountain permafrost are not extensive and monitoring is confined essentially to the European Alps. Evidence from borehole profiles in permafrost helps to determine the rate and magnitude of temperature changes (Mackay, 1990, 1992; Vonder Mühll and Holub, 1995). Vonder Mühll *et al.* (1994) have shown that permafrost temperatures are increasing at an annual rate of 0.1°C; this reflects the particularly strong warming signal in the Alps since the 1980s. It is difficult to interpret changes in permafrost, as the amount of permafrost in any particular region depends on regional geomorphological characteristics, soil and geology, exposure to weather elements, seasonal precipitation and temperature, and the amount and duration of the snow-pack. This latter parameter significantly controls perennially frozen surfaces, because snow insulates the ground surface and suppresses the propagation of cold temperatures into the soil which are favourable to permafrost. Changes in just one of the controls on permafrost, such as temperature, need to be viewed in the larger context of the other controls on this element of the cryosphere.

6.4 Extreme events and their impacts on geomorphological features

One of the numerous problems associated with projections of climate change in the next century under conditions of enhanced atmospheric greenhouse gas concentrations is the manner in which the variability and the extreme values of temperature and precipitation may change (Beniston, 1994, 1997; Rebetez and Beniston, 1998). In terms of the response of a number of environmental and socio-economic systems, extreme or highly variable climatological events tend to have a greater impact than changes in mean climate, which is commonly the basic 'quantity' for discussing climatic change. Katz and Brown (1992) have shown that the frequency of extreme events is more sensitive to variability than to averages.

There is speculation that a warmer climate will induce more extremes of storms, floods, droughts and heatwaves, though this is by no means certain (IPCC, 1996). Mearns *et al.* (1994) have shown that the changes in the probabilities of extreme events and those in the corresponding mean temperatures are quite non-linear. The latitude and altitude of different mountain systems determine the relative amount of snow and ice at high elevations and intense rainfall at lower elevations. In a changed climate, meteorological extremes may also occur in regions that are today less prone to such events, and vice versa. Because of the amount of precipitation and relief, and the fact that many mountain ranges are located in seismically active regions, the added effect of intense rainfall in low- to middle-altitude regions is to produce some of the highest global rates of slope erosion. Climate change could alter the magnitude and/or frequency of a wide range of geomorphological processes (Eybergen and Imeson, 1989). Increases in extreme precipitation events, associated with

snow-melt, could increase the frequency and severity of floods. In the mountain regions of Asia affected by the monsoon, for example, loss of revenue in agriculture and damage to infrastructure related to monsoon storms represent a high proportion of annual loss, as reported by Carver and Nakarmi (1995). Increased severity and frequency of storms during the monsoon period in the future will lead to higher rates of erosion and more frequency of flooding in the Himalayas (Yoshino *et al.*, 1998). Such extreme events would affect erosion, discharge and sedimentation rates, which damage hydro-power infrastructure; furthermore, sediments deposited in large quantities on agricultural lands, irrigation canals and streams would lead to reductions in agricultural production.

Large rockfalls in high mountainous areas are often caused by groundwater seeping through joints in the rocks. If average and extreme precipitation were to increase, then groundwater pressure would rise, providing conditions favourable to increased triggering

---

Box 6.2   Debris flow and climate change in Switzerland

Rebetez *et al.* (1997) have undertaken an exhaustive study of debris flows in the Ritigraben region of the Valais Alps of Switzerland. These flows, which generally occur in late summer and early fall, have been analysed in relation to meteorological and climatic factors. The principal trigger mechanisms for debris flows are abundant rain on the one hand, and snow-melt and runoff on the other hand, or a combination of both. Debris flows have been observed to occur when total rainfall amount over a three-day period exceeds four standard deviations, representing a significant extreme precipitation event.

An analysis of climatological data for the past three decades in the region of Ritigraben has highlighted the fact that the number of extreme rainfall events capable of triggering debris flows in

August and September has increased. Similar trends are observed for the twentieth century elsewhere in Switzerland, implying a link between observed warming and increases in extreme precipitation events.

The general temperature rise in a region of permafrost may also play a role in the response of slope stability to extreme precipitation. At the foot of the Ritigraben, warming trends of both minimum and maximum temperatures have been particularly marked in the past two decades.

This situation is likely to increase in severity in a warmer climate as projected by regional climate simulations for the Swiss region (Beniston *et al.*, 1995). The consequences for damage to human infrastructure in the valley below the Ritigraben, including the road and rail links to the resort of Zermatt, would represent a severe financial burden for the local mountain communities.

of rockfalls and landslides. Large landslides are propagated by increasing long-term rainfall, whereas small slides are triggered by high-intensity rainfall (Govi, 1990). In a future climate, where both the mean and the extremes of precipitation may rise in certain areas, there would be a corresponding increase in the number of small and large slides. This could in turn contribute to additional transport of sediments in the river systems originating in mountain regions. Other trigger mechanisms for rockfalls are linked to pressure-release joints following deglaciation (Bjerrum and Jfrstad, 1968); such rockfalls may be observed decades after the deglaciation itself, emphasizing the long time lags involved. Freeze–thaw processes are also very important (Rapp, 1960), and several authors have reported possible links between rockfall activity and freeze–thaw mechanisms linked to climate change (Senarclens-Grancy, 1958; Heuberger, 1966).

A further mechanism responsible for decreased slope stability in a warmer climate is the reduced cohesion of the soil through permafrost degradation (Haeberli *et al.*, 1990). With the melting of the present permafrost zones at high mountain elevations, rock- and mudslide events can be expected to increase in number and, possibly, in severity (Iafiazova, 1997). This will certainly have a number of economic consequences for mountain communities, where the costs of repair to damaged communications infrastructure and buildings will rise in proportion to the number of landslide events.

In many mountainous regions, tourist resorts such as those in the Alps and the Rocky Mountains, or large urban areas close to mountains (suburbs of South American Andean cities, Hong Kong or Los Angeles, for example), have spread into high-risk areas. These will be increasingly endangered by reductions in slope stability.

6.5 Impacts on ecological systems

6.5.1 Vegetation

Biological systems follow a hierarchy which includes biomes, ecosystems, communities,

species populations and even isolated individuals. At the individual level, the chain can be extended to organs, tissues, cells and genes (which are considered to be macromolecules; Guisan *et al.*, 1995b). Climate change can impact on any one part of the hierarchy levels or simultaneously at all levels. Because of the complexity of biological systems, attempting to disaggregate the most important biological elements' response to environmental change is a major challenge. Körner (1993) suggests that assessing impacts at the ecosystem level may be the most relevant, since ecosystem changes result from interactions at both the lower and the higher complexity levels.

Plant life at high elevations is primarily constrained by direct and indirect effects of low temperatures and by the reduced quantities of available CO_2 compared to lowland regions. Other climatic influences, such as radiation, wind and storminess or insufficient water availability, may be important regionally (Fliri, 1975; Lauscher, 1977; Körner and Larcher, 1988). Plants respond to these climatological influences through a number of morphological and physiological adjustments such as stunted growth forms and small leaves, low thermal requirements for basic life functions, and reproductive strategies that avoid the risk associated with early life phases.

Because temperature decreases with altitude by 0.5–1.0°C/km, a first-order approximation concerning the response of vegetation to climate change is that species will migrate upwards to find climatic conditions in tomorrow's climate which are similar to today's (e.g. MacArthur, 1972; Peters and Darling, 1985). According to this hypothesis, the expected impacts of climate change in mountainous nature reserves would include the loss of the coolest climatic zones at the peaks of the mountains and the linear shift of all remaining vegetation belts upslope. Because mountain tops are smaller than bases, the present belts at high elevations would occupy smaller and smaller areas, and the corresponding species would have smaller populations and might thus become more vulnerable to genetic and environmental pressure (Peters and Darling, 1985; Hansen-Bristow *et al.*, 1988; Bortenschlager, 1993).

While temperature is indeed a major controlling factor on vegetation, it is by no means the only one, nor is it necessarily the dominant factor for some species. Certain types of vegetation have a large tolerance range to temperature, while others may suffer severe damage in response to a relatively modest change in climatic conditions. Some plants are capable of migrating by various seeding mechanisms, while others can adapt within their original environment. Plant communities will experience competition, with the most robust species adapting or migrating at the expense of the less robust ones. Furthermore, species capable of migrating upwards may not find other adequate environmental conditions, in particular soil types and soil moisture, such that even if climate is appropriate at a particular site in the future, the plant's migration will be slowed or suppressed by other factors. In populated mountain regions of the world, additional stress factors include the fragmentation of biomes and the obstacles to migration, such as roads and settlements. The success of colonization may also depend on regions where erosion and overland flows may increase under changing climatic conditions. Whatever the response of ecosystems, it must be borne in mind that they are exceedingly complex dynamic systems whose response is likely to be non-stationary and stochastic (Hengeveld, 1990). The ability for species to maintain viable populations at ecoclines or ecotones will be affected by numerous interactions between existing populations and site-specific factors (Halpin, 1994). These interactions will be contained in a complex cascade of environmental and ecological feedbacks.

Huntley (1991) suggests that there are three responses which can be distinguished at the species level, namely genetic adaptations, biological invasions through species inter-competition, and species extinction. Street and Semenov (1990) show that these responses may take one of the following adaptation pathways in the face of changing environmental conditions. A first scenario is that the currently dominant species would progressively be replaced by a more thermophilous (heat-loving) species. A second plausible mechanism is that the dominant species is replaced by pioneer species of the same community that have enhanced adaptation capability. A third possibility is that environmental change may favour less dominant species, which then replace the dominant species through competition. These scenarios are based on the assumption that other limiting factors such as soil type or moisture will remain relatively unaffected by a changing environment.

Halpin (1994) has used geographical information systems (GIS) techniques to map the changes in vegetation that could take place under hypothetical climate change scenarios. By mapping the potential zones of shifts in thermal belts with altitude, and knowing the potential for one species to dominate over another, he has applied the concept to three mountain regions of the western hemisphere, namely Costa Rica, the Sierra Nevada of California, and the Alaskan ranges. In Figure 6.3, an example is given of the shifts in vegetation in a changed climate in the Californian mountains. It is seen that extinction is projected to occur for the montane desert scrub and the subalpine dry scrub as a result of species inter-competition. While reduced in area, the alpine tundra zone located towards the tops of the mountains is able to survive under this particular scenario. Results for the other ranges suggest species extinction at the mountain tops in Costa Rica, with little change occurring at height in Alaska. Examples of past extinctions attributed to upward shifts are found in Central and South America, where vegetation zones have moved upward by 1000–1500 m since the last glacial maximum (van der Hammen, 1984; Flenley, 1979). Romme and Turner (1991), in their study on possible implications of climate change for ecosystems in Yellowstone National Park (USA), project species extinctions as a result of fragmentation and shrinking mountain-top habitats.

The examples provided by the Halpin (1994) study emphasize the caution which is needed when interpreting such results; even a rather simple conceptual model suggests that

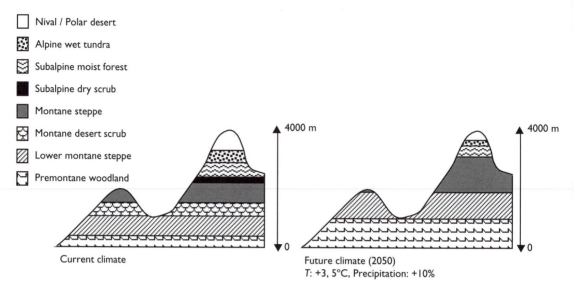

☐ Nival / Polar desert

▨ Alpine wet tundra

▨ Subalpine moist forest

■ Subalpine dry scrub

▨ Montane steppe

▨ Montane desert scrub

▨ Lower montane steppe

▨ Premontane woodland

Current climate

Future climate (2050)
T: +3, 5°C, Precipitation: +10%

FIGURE 6.3 Example of vegetation changes under different climatic conditions, as a function of the movement of climatic ranges along altitudinal gradients and intercompetition between vegetation communities
Source: After Halpin (1994)

changes in ecoclimatic zones are not linear functions of altitudinal climatic gradients.

It is expected that, on a general level, the response of ecosystems in mountain regions will be most important at ecoclines (the ecosystem boundaries if these are gradual) or ecotones (where step-like changes in vegetation types occur). Guisan *et al.* (1995b) note that ecological changes at ecoclines or ecotones will be amplified because changes within adjacent ecosystems are juxtaposed. In steep and rugged topography, ecotones and ecoclines increase in quantity but decrease in area and tend to become more fragmented as local site conditions determine the nature of individual ecosystems. Even though the timber line is in many regions not a perfect ecocline, it is an example of a visible ecological boundary which will be subject to change in coming decades. This change will take place either in response to a warmer climate, or as a result of recolonization of pastures that have been cleared in the past for pastoral activities.

McNeely (1990) has suggested that the most vulnerable species at the interface between two ecosystems will be those which are genetically poorly adapted to rapid environmental change. Those which reproduce slowly and disperse poorly, are isolated or are highly specialized will therefore be highly sensitive to seemingly minor stresses.

Numerous investigations have attempted to show that negative effects of climatic change may be offset by the enhanced atmospheric concentrations of CO_2 in the future. Körner (1995) has reviewed this problem and notes that on average, high-altitude plants are able to fix more CO_2 per unit leaf area than are lowland plants. However, the net uptake is about the same for both high- and low-elevation plants because of the lower CO_2 levels at height. In an atmosphere with enhanced CO_2, therefore, high-altitude plants are expected to improve their primary productivity compared to today. Körner *et al.* (1995) have used experimental plots near the Furka Pass in central Switzerland to detect changes in alpine plant productivity as a result of CO_2 enrichment. The plants subjected to enrichment were seen to have a higher photosynthetic rate per unit area compared with control plants at current CO_2 levels. Other studies have also shown this acclimation effect, which in certain species diminishes over time (e.g. Gunderson and

Wullschleger, 1994). However, these experiments under carefully controlled conditions are probably far removed from changes which may take place in the free environment in coming decades; even if photosynthetic uptake of carbon by alpine plants were to be the rule, the consequences for plant growth and development are still uncertain.

Callaghan and Jonasson (1994) have shown that the direct effects of higher temperatures exert greater controls on above-ground plant processes, compared to the indirect effects of warming on processes taking place within the soil. Furthermore, such processes will be enhanced in regions where climatic conditions are currently extreme, as indicated by Bugmann and Fischlin (1994). In regions where climatic change may lead to warmer and drier conditions, mountain vegetation could suffer as a result of increased evapotranspiration. This is most likely for mountains in climates typical of continental and Mediterranean regimes, although even in tropical regions there are indications that plants are already sensitive to water stress on mountains such as Mt Kinabalu in the Malaysian part of Borneo (Kitayama, 1996).

The length and depth of snow cover, often correlated with mean temperature and precipitation, is one of the key climatic factors in alpine ecosystems (Barry and Van Wie, 1974; Aulitzky *et al.*, 1982; Ozenda, 1985; Burrows, 1990). Snow cover provides frost protection for plants in winter, and water supply in spring. Alpine plant communities are characterized by a very short growing season (i.e. the snow-free period) and require water to commence growth. Ozenda and Borel (1991) have predicted that vegetation communities which live in snow beds and in hollows will be the most vulnerable to change, because they will be subject to summer desiccation.

There are currently a number of ecosystem models available which can be used to test the sensitivity of a particular system to processes such as nutrient cycling (e.g. CENTURY), for investigating species composition under changed environmental conditions (e.g. BIOME, DOLY, MAPSS; Woodward *et al.*, 1995), or for assessing forest health (e.g.

FORET; Innes, 1998). The problem here is that many of these models are nominally capable of operating at higher spatial resolution than the GCMs which are providing part of the essential input data, i.e. spatially distributed climate scenario data. A supplementary problem is that these models are designed to operate over continental to global scales, with little possibility of emphasis on the details of altitudinal vegetation-belt typology and dynamics typical of mountain regions. Ecosystems are expected to shift upwards and polewards in response to global warming, but expansion will be determined by edaphic factors and dispersal rates, which will determine colonization rates. Additional constraints include land-use competition by human activities in many parts of the world, including the Middle East, where significant biodiversity is being lost as a result of human encroachment (settlements; industry; flooding of valleys by hydro-power facilities; etc.), as reported by Bie and Imevbore (1995) and Kharin (1995). In the Himalayas, Yoshino *et al.* (1998) suggest that weedy species with a wide ecological tolerance will be able to adapt more than other species to shifts in temperature and precipitation. Mountain forests in some regions of South-East Asia, Africa and Central and South America are presently under greater threat from direct human interference (deforestation; slash-and-burn practices) than from climatic change *per se*; unless there is a reversal of current deforestation trends, this situation is unlikely to change in the near future. Sinha and Swaminathan (1991) believe that the combined effects of deforestation and climatic change may impact heavily on sustainable food security, and of course on native populations who still dwell in isolated forested mountain regions.

A number of modelling studies have been conducted using forest gap models (Shugart, 1984) to assess the impacts of climatic change on forest biomass and species composition in mountainous regions (e.g. Kienast, 1991; Kräuchi and Kienast, 1993; Bugmann, 1994; Bugmann and Fischlin, 1994; Kräuchi, 1994). Despite different climate scenarios, similar tendencies were found in terms of forest response to change in the European Alps. Here, the main climatic space contraction and fragmentation of plant populations would be in the present alpine and nival belts, where rare and endemic species with low dispersal capacities could become extinct. As the alpine belt contains about 15 per cent of the endemic alpine flora (Ozenda and Borel, 1991), the potential impact of climate change on floristic diversity in the Alps could be significant (Grabherr *et al.*, 1994).

Model studies for Australian mountain vegetation show that both trees and shrubs may invade current mountain pastures in response to rising temperatures (Williams and Costin, 1994). Tree line may rise by 100 m for each 1°C increase in mean annual temperature (Galloway, 1988). Elevated summer temperatures may also lead to an expansion of shrub communities (Williams, 1990). However, because of the relatively small surface area of the more elevated parts of Australia's mountain ranges, there is a distinct possibility that certain species already located close to the tops of the mountains may become extinct. This would particularly involve those which cannot adapt in their present environment, because they have no migration potential. This is indeed the case in many mountain regions of the world, where vegetation is today confined to mountain tops and therefore highly fragmented in space.

VEMAP, a continental-scale vegetation response study of the United States, considered how three biogeographical models (BIOME2, DOLY, MAPSS) respond to a double-CO_2 scenario. Simulated alpine and subalpine regions in the western United States migrate to higher elevations, and thus decrease in area, while subalpine montane forest boundaries also move upward (Woodward *et al.*, 1995). Using gap model simulations applied to British Columbian mountains, Cumming and Burton (1996) have shown that certain upward-moving forest ecosystems could actually disappear from their potential habitats because of the lack of winter cooling, vital for regeneration and the robustness of trees, and a greater sensitivity to droughts and frosts. In all forest impact studies, in both

latitudinal and altitudinal terms, climatic change as projected by the IPCC (1996) will be more rapid than the migrational capacity of forests. The faster the rate of environmental change, the greater the probability of species extinction and the disruption of ecosystems. Halpin (1994) has shown that migration processes are exceedingly complex. Even if vegetation belts do not move up as a whole in response to global climate change, the ecological potential of sites will change in relation to profound changes in their environment.

It should be emphasized that there are considerable limitations in present-day simulation techniques for assessing ecosystem response to climate change, in particular the temporal changes of these responses. In general, increases in atmospheric temperature will affect the structure and function of vegetation, and also species composition where time may not be sufficient to allow species to migrate to suitable habitats (Kienast, 1991; Bugmann, 1994; Klötzli, 1994). According to the detail of biogeographical models, including for example the response of vegetation not only to temperature but also to the CO_2 fertilization effect, results can be very different and sometimes even contradictory. Shriner and Street (1998) have shown that without the CO_2 effect and with a moderate increase in mean temperatures, forests respond by increased growth, while the reverse is frequently observed in simulations where the fertilization effect is absent. On the other hand, even when CO_2 influences are taken into account, forest dieback becomes significant when global warming is towards the upper range of IPCC scenarios. A long-term consequence of global change could comprise two phases, one in which enhanced growth occurs as a first response to warming, followed by dieback when warming exceeds a particular threshold beyond which particular forest species can no longer survive.

The impact of climatic change on altitudinal distribution of vegetation cannot be analysed without taking into account interference with latitudinal distribution. Especially at low altitudes, Mediterranean tree species can substitute for submontane belt species. In the southern French Alps, Ozenda and Borel (1991) predict a northward progression of Mediterranean ecotypes ('steppification' of ecosystems) under higher temperatures and lower rainfall amounts. Kienast *et al.* (1998) have applied a spatially explicit static vegetation model to alpine vegetation communities. The model suggests that forests which are distributed in regions with low precipitation and on soils with low water-storage capacity are highly sensitive to shifts in climate. Under conditions of global warming, the northward

progression of Mediterranean influences would probably be important, and it is estimated that 2–5 per cent of currently forested areas of Switzerland could undergo 'steppification', particularly on the Italian side of the Alps and in the intra-alpine dry valleys. A similar change is less likely to take place in the south-eastern part of the range (the Julian and Carnic Alps), where the climate is much more humid. In boreal latitudes, migration of the tree line polewards into previously barren regions would significantly modify the surface

---

Box 6.3 The Australian pygmy-possum: towards a species extinction?

The pygmy-possum is the smallest member of the marsupial family in Australia. It is an extremely rare species, estimated at about 2500 individuals (Walter and Broome, 1998), whose habitat is found within the Australian Alps in the south-eastern part of the continent. These small mammals are confined to the higher elevations above 1400 m, which corresponds to the lowest limit of the winter snow line. Snow cover and duration is an important element of the animals' metabolism. The pygmy-possums are a hibernating species, and the melting of the snow-pack in spring corresponds to the start of the reproductive cycle. These animals are therefore genetically adapted to snow, which insulates their winter hideouts from winter temperatures; even if the variability of snow in Australia is large from one year to the next (Danks, 1991), there is always a minimum amount of snow in the upper reaches of the mountains which allows normal hibernation. Because they have a relatively low tolerance to heat and other environmental factors, the pygmy-possums do not venture outside their climatic range.

Whetton *et al.* (1996) have analysed climate model results and project a substantial increase in temperatures and decrease in the duration of the snow season. As a result, the conditions for adequate hibernation will

diminish. Because the pygmy-possums are so dependent on ambient temperature for energy expenditure (Geiser and Broome, 1993), and because the current habitat distribution is so closely linked to snow, any changes in these conditions in a warmer global climate are likely to be detrimental to the survival of the species. Because the altitudinal range of the Australian mountains is relatively modest, the pygmy-possums have practically no possibilities for migration. It is thus likely that the pygmy-possum will become extinct even with a modest 1°C average rise in temperature (Bennett *et al.*, 1991).

In theory, measures could be taken to assist the pygmy-possum in order to avert or to postpone ultimate extinction, as climate and environmental conditions become increasingly unfavourable to these animals. For example, a zoo-type enclosure with an artificial climate could technically be feasible. Whether the pygmy-possums would in fact be capable of adapting to a constrained environment is highly uncertain; in addition, within what time-frame would one wish to avert extinction, i.e. would extinction be in any case inevitable (either through non-intervention or through captivity)? While a species extinction as a result of indirect human interference may find short-term technological solutions, the longer-term prospects need to find a basis in terms of science and ethics.

characteristics and local climates, in particular through changes in albedo and surface energy balance (Fitzharris *et al.*, 1996). With the expansion of boreal forest zones in both mountain and lowland regions, new assemblages of plant and animal species can be expected in regions such as the northern Alaskan ranges and the eastern Siberian mountains.

It has often been suggested that currently protected areas, such as national parks, could become biological refugia for plant and animal species under rapidly changing environmental and climatic conditions. Conservation of biodiversity could be achieved through a reduction of ecosystem fragmentation, anthropogenic stresses, and by increasing the connectivity among habitats, e.g. 'migration corridors' (Markham and Malcolm, 1996). Refugia would thus theoretically play the role of allowing ecosystems to adapt or migrate in the absence of human interference. However, Kienast *et al.* (1998) have shown, on the basis of a vegetation modelling study, that for the Swiss National Park, almost 50 per cent of the migrating species would be unable to be hosted because of the limited vertical extension of the park. This is likely to be a general problem for other regions as well. Refugia are likely to be themselves fragmented, and therefore the possibility of allowing migration corridors for many species would be extremely limited, other than in regions which are still relatively isolated and far removed from significant human activity.

6.5.2 Soils

Soils can be affected in a number of ways by environmental change. Under more extreme conditions (precipitation or storminess), soils might be washed or blown away at faster rates than today; this would in turn affect the ecosystem composition and could lead to the disappearance of species if the vital soils were to disappear completely from certain mountain slopes. Tropical and monsoon regions may be particularly at risk to enhanced landslides, mudslides and flash flooding, but even in subhumid mountain environments, heavy rain following a prolonged drought or

fire might lead to irreversible damage to landscape, as vegetation might no longer be able to recover and restabilize slopes.

Soil chemistry may also be significantly affected by climatic change, through enhanced or reduced leaching; leaching would be particularly enhanced in certain tropical regions. More elevated temperatures will generally increase the rate of organic matter decomposition within the soils, although this could to some extent be offset by the stimulation of the primary production of plants under future CO_2 conditions (McMurtrie and Comins, 1996; Thornley and Cannell, 1996). Modifications of nitrate, carbon and other organic matter content of the soils, and their relative proportions, could become important where vegetation recycling is enhanced by warmer temperatures and higher atmospheric CO_2 levels. In regions where rainfall may diminish, water availability is likely to be a severe limiting factor and could lead to desertification of currently marginal mountain regions (Le Houerou, 1989; Kassas, 1995).

It should be emphasized that soil development is slow, and especially so at high elevations; as a consequence, changes in soil composition will lag vegetation changes, which themselves will occur slowly in response to abrupt shifts in temperature and moisture conditions.

6.5.3 Fire

Fire is an element which is of particular importance in many ecological systems; it can be devastatingly destructive in certain circumstances, but it plays a vital role in the recycling of organic material and the regrowth of vegetation. Changing climatic conditions are likely to modify the frequency of fire outbreaks and intensity, but other factors may also play a major role; for example, changes in fire-management practices and forest dieback may lead to a weakening of the trees in response to external stress factors (Fosberg, 1990; King and Neilson, 1992). In North America in particular, fire management favoured suppression of forest fires in recent decades, and as a consequence there has been a substantial increase in biomass compared to natural levels. Under such circumstances, Stocks (1993) and Neilson *et al.* (1992) have shown that forests tend to transpire most of the available soil moisture, so that catastrophic fires can occur as a result of the greater sensitivity of trees to seemingly minor changes in environmental conditions. One example of the combination of deadwood accumulation resulting from fire-suppression policies and a prolonged drought is the long and spectacular fire outbreak which occurred in Yellowstone National Park in the United States during the summer of 1988. Fires can thus occur in the absence of significant climate change.

With climatic change as projected by the IPCC (1996), prolonged periods of summer drought would transform areas already sensitive to fire into regions of sustained fire hazard. The coastal ranges of California, the Blue Mountains of New South Wales (Australia), Mt Kenya and mountains on the fringes of the Mediterranean Sea, already subject to frequent fire episodes, would be severely affected. Fires are also expected to occur in regions which are currently relatively unaffected, as critical climatic, environmental and biological thresholds for fire outbreaks are exceeded.

In regions which currently experience slash-and-burn agriculture, the natural vegetation cover is rapidly changing, in terms of the structure, species composition and age diversity of the trees. It is difficult to evaluate the manner in which fire frequency may change in areas where the natural vegetation cover is rapidly being replaced by other species.

Because many regions sensitive to fires are located close to major population centres, there could be considerable damage to infrastructure and disturbances to economic activities at the boundaries of cities such as Los Angeles and the San Francisco Bay Area in California, Sydney in Australia, and coastal resorts close to the mountains in Spain, Italy and southern France. This has already occurred in the past and is likely to become more frequent in the future as fire hazards increase and urban centres expand as a result of population growth.

7

Socio-economic impacts of environmental change in mountain regions

7.0 Chapter summary

As a follow-on from Chapter 6, this chapter discusses the potential socio-economic responses to environmental changes in mountains and uplands. It is often necessary to distinguish between the social and economic responses to change in the developing world and in the industrialized countries respectively, because of the vastly different significance of mountains for indigenous populations. Particular sectors which are focused upon in this chapter include mountain agriculture, human health, tourism, energy and other commercial activities. For each of these sectors, it is seen that mountains are not simply marginal in the resources which they provide, but have a significant economic bearing for numerous countries. Environmental change may upset the delicate balance which has in many regions been forged over centuries between human populations and their environment, leading to economic and social hardship. The development of new economic resources, such as tourism, poses the problem of economic promotion versus environmental protection.

7.1 Introduction

Few assessments of the impacts of environmental change on socio-economic systems have been conducted in mountain regions, in contrast to other regions such as coastal zones, or arid and semi-arid areas. A number of reasons can be put forward for this situation, such as the fact that many mountains and uplands are economically and politically marginal. In individual countries, their direct economic importance to national economies is generally low, and they are often far from centres of political decision-making.

The complexity of mountain systems presents major problems for assessing the potential impacts of environmental change. This applies to assessments of changes in both biophysical systems (e.g. Rizzo and Wiken, 1992; Halpin, 1994) and societal systems, particularly because many of the most valuable products of mountain regions are not easily quantified in monetary terms (Price, 1990). Tourism, which is an increasingly important component of mountain economies around the world, represents an economic sector which, in contrast to agriculture or forestry, is closely linked to other aspects of mountain economies and is not a 'stand-alone' sector. A further consideration is that few mountain regions can be described comprehensively, because of the heterogeneity of the available data for climatic, biological or socio-economic impact assessments.

Because of the diversity of mountain economies, from the exclusively tourist-based ones to those characterized by subsistence agriculture, no single impacts study will adequately represent the range of potential socio-economic responses to climate change. A case-by-case approach is therefore essential to

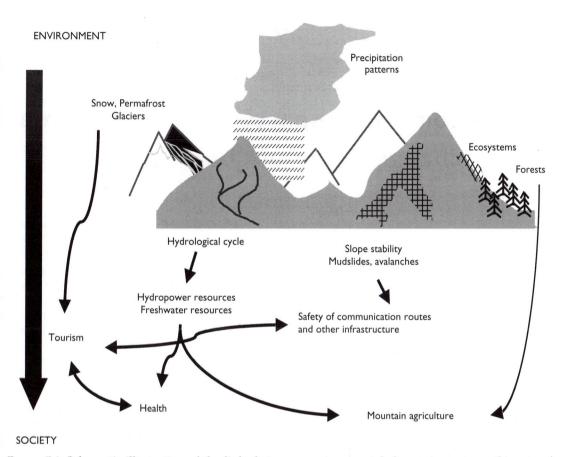

ENVIRONMENT

Precipitation patterns

Snow, Permafrost Glaciers

Ecosystems

Forests

Hydrological cycle

Slope stability
Mudslides, avalanches

Hydropower resources
Freshwater resources

Safety of communication routes
and other infrastructure

Tourism

Health

Mountain agriculture

SOCIETY

FIGURE 7.1 Schematic illustration of the links between environmental change impacts on the natural environment and the repercussions of these on socio-economic systems

understand how, for example, mountain agriculture may change in Bolivia, tourism may change in Switzerland, or hydro-power resources may be impacted in New Zealand.

In view of the fact that humans have influenced mountain ecosystems in many different ways throughout history, anthropogenic impacts generally cannot be dissociated from climate change impacts, as illustrated schematically in Figure 7.1. Climatic influences are often obscured by the impacts of change in land use. An example is the fragmentation of the forest and natural vegetation cover. Because of persistent anthropogenic influences in the past, the timber line in mountains such as the Alps has dropped between 150 and 400 m compared to its uppermost position during the post-glacial optimum (Holtmeier,

1994). At present, the climatic limit of tree growth in the Alps is situated above the actual forest limit (Thinon, 1992; Tessier *et al.*, 1993). By reducing species diversity and even intra-species genetic variability of some species, humans have reduced the ability of alpine vegetation to respond to environmental change (David, 1993; Peterson, 1994).

In general terms, it should be emphasized that many sectors of global environmental change will impact upon the poorest members of society, who are clearly the most vulnerable to changes which will affect mountains in coming decades. The potential impacts of global change will probably exacerbate hunger and poverty around the world. New and fluctuating conditions could have a strongly negative impact on economic activities, particularly in the resources sector. People who are highly dependent on farming and forestry might see their livelihood severely disrupted by changes in rainfall patterns, impoverished forests and degraded soils. The poor would suffer the most because they have fewer options for responding to environmental change, in terms of technological and financial resources. For example, they would find it more difficult to change over to new crops requiring less water, to pump water for irrigation, or to extend their cultivatable land. Such solutions typically require extensive inputs such as machinery or fossil fuel energy, which are beyond the financial capabilities of the populations concerned. If global change were to have severe local or regional impacts in certain mountains and uplands, then waves of refugees and immigrants would be likely to move from rural to urban areas within national borders, and from the South to the North across national boundaries. Such migrations from non-urban populations would probably become an additional source of social and political conflict. Displaced and impoverished populations would suffer an erosion of their cultural identity. The resulting disruption to their culture might create social and political problems just as intractable as the environmental problems that generated them in the first place (IUC, 1997).

This chapter will consider four socio-economic sectors of high relevance to mountain regions, namely mountain agriculture, health, tourism and energy.

7.2 Mountain agriculture

Upland regions contribute a significant proportion of the world's agricultural production in terms of economic value. While mountainous areas in the middle and high latitudes are often marginal for agricultural production compared with the warmer lowland areas, the converse may be true at lower latitudes, where highland zones frequently offer a more temperate climate. In addition, because mountains and uplands are the source region for many of the world's major rivers, changes in environmental conditions may modify the seasonal character and the amount of discharge in hydrological basins. This will in turn disrupt the lowland agriculture that is dependent on the availability of water in these rivers.

Upland regions are characterized by climatic gradients that can lead to rapid altitudinal changes in agricultural potential over comparatively short horizontal distances. Where elevations are high enough, a level will eventually be reached where agricultural production ceases to be viable, in terms of either economic profit or subsistence. Upland crop production, practised close to the margins of viable production, can be highly sensitive to variations in climate. The nature of that sensitivity varies according to the region, crop and agricultural system of interest. In some cases, the limits to crop cultivation appear to be closely related to levels of economic return. Yield variability often increases at higher elevations, so that climate change may mean a greater risk of yield shortfall, rather than a change in mean yield (Carter and Parry, 1994).

The impacts of climatic change on agriculture can be assessed at different scales including crop yield, farm or sector profitability, regional economic activity or hunger vulnerability. Impacts depend on biophysical and socio-economic responses. The methodological approaches may include different levels of interaction with related sectors.

The world food system involves a complex dynamic interaction of producers and consumers, interlinked through global markets. Related activities include input production and acquisition, transportation, storage and processing. While there is a trend towards internationalization in the world food system, only about 15 per cent of total world production currently crosses national borders (Fischer *et al.*, 1990). National governments shape the system by imposing regulations and through investments in agricultural research, infrastructure improvements and education. The system functions to meet the demands for food, to produce food in increasingly efficient ways, and to trade food within and across national borders. Although the system does not guarantee stability, it has generated long-term real declines in prices of major food staples (ibid.).

The effects of climate on agriculture in individual countries cannot be considered in isolation. Agricultural trade has grown in recent decades and now provides significant increments of national food supplies to major importing nations and substantial income for major exporting ones. There are therefore close links between agriculture and climate, the international nature of food trade and food security, and the need to consider the impacts of climate change in a global context. Despite technological advances such as improved crop varieties and irrigation systems, weather and climate are still key factors in agricultural productivity. For example, weak monsoon rains in 1987 caused large shortfalls in crop production in India, Bangladesh and Pakistan, which forced these countries to import corn (World Food Institute, 1988).

Recent research has focused on regional and national assessments of the potential effects of climate change on agriculture. At the same time, there has been a growing emphasis on understanding the interactions of climatic, environmental and social factors in a wider context (Parry *et al.*, 1990). Relatively little work has been systematically undertaken, however, to identify vulnerable socio-economic groups, integrate effects across sectors, describe impacts at different spatial and temporal scales, or address the efficacy of the range of practicable responses (Parry *et al.*, 1992). Many studies have treated each region or nation without relation to changes in production in other places, and have not addressed in an integrated way the interactions with other related environmental sectors, such as water resources.

Agricultural production will be affected by the severity and pace of climate change. If change is gradual, there will be time for political and social institutions to adjust. Slow change also may enable natural biota to adapt. Many untested assumptions lie behind efforts to project global warming's potential influence on crops. In addition to the magnitude and rate of change, the stage of growth during which a crop is exposed to drought or heat is important. When a crop is flowering or fruiting, it is extremely sensitive to changes in temperature and moisture; during other stages of the growth cycle, plants are more resilient. Moreover, temperature and seasonal rainfall patterns vary from year to year and region to region, regardless of long-term trends in climate. Temperature and rainfall changes induced by climate change will most likely interact with atmospheric gases, fertilizers, insects, plant pathogens, weeds and the soil's organic matter to produce unanticipated responses. Rainfall is the major limiting factor in the growth and production of crops worldwide. Adequate moisture is critical for plants, especially during germination and fruit development.

As agricultural production derives from various types of plants and can be described in terms of both amount and quality, the reactions of an individual crop to environmental change will depend on the balance of shorter cycles, shorter periods to accumulate yield products, higher potential yields due to increased assimilation of CO_2, and increased water-use efficiency due to enhanced regulation of transpiration with elevated CO_2. Therefore, increasing carbon dioxide levels in the atmosphere should improve the rates of growth and water use among many crops (Goudriaan and Unworth, 1990). Most plants growing in experimental environments with increased levels of atmospheric CO_2 exhibit increased rates of net photosynthesis (i.e. total photosynthesis minus respiration) and reduced stomatal openings. Experimental effects of CO_2 on crops have been reviewed by Acock and Allen (1985) and Cure (1985). Partial stomatal closure leads to reduced transpiration per unit leaf area and, combined with enhanced photosynthesis, often improves water-use efficiency (the ratio of crop biomass accumulation or yield to the amount of water used in evapotranspiration). Thus, by itself, increased CO_2 can increase yield and reduce water use per unit biomass. On the basis of laboratory studies, doubling pre-industrial carbon dioxide levels should increase crop yields significantly.

Projections of future agricultural production stem from both 'controlled conditions' experiments and from initializing crop simulation models with climate scenario data. Quantitative results of simulations are therefore highly dependent on the type of climate scenario used, especially in terms of shifts in precipitation regimes. Furthermore, most simulation and experimental studies have so far used expected fluctuations of mean values for climate variables, but the emphasis is increasingly shifting to the possible consequences of a more variable interannual and intra-annual climate and extremes.

Several authors have predicted that currently viable areas of crop production will change as a result of climate change in the Alps (Balteanu *et al.*, 1987), Japan (Yoshino *et al.*, 1988), New Zealand (Salinger *et al.*, 1989) and Kenya (Downing, 1992). These projections do not consider other constraints such as soil types which may no longer be suitable for agriculture at higher elevations. In-depth studies of the effects of climatic change in Ecuador's Central Sierra (Parry, 1978; Bravo *et al.*, 1988) and Papua New Guinea (Allen *et al.*, 1989) have shown that crop growth and yield are controlled by complex interactions between various climatic factors. Specific methods of cultivation may permit crop survival in sites where the microclimates would otherwise be unsuitable. Future climate scenarios suggest both positive and negative impacts, such as decreasing frost risks in the Mexican highlands (Liverman and O'Brien, 1991) and less productive upland agriculture in Asian mountains, where impacts would depend on various factors, particularly types of cultivars and the availability of irrigation (Parry *et al.*, 1992).

Given the wide range of microclimates already existing in mountain areas and which have been exploited through cultivation of diverse crops, the direct negative effects of climate change on crop yields may be relatively small. While crop yields may rise if moisture is not limiting, increases in the number of extreme events may offset any potential benefits. In addition, increases in both crop and animal yields may be negated by greater populations of pests and disease-causing organisms, many of which have distributions that are climatically controlled. Global warming will favour conditions for insects to multiply and prosper. Rising temperatures will lengthen the breeding season and increase the reproductive rate. This in turn will raise the total number of insects attacking a crop and subsequently increase crop losses. In addition, some insects will be able to extend their range northward as a result of the warming trend (Chippendale, 1979).

Weeds which are better adapted to arid conditions than crops will provide increased competition for moisture, nutrients and light. Herbicidal controls are less effective under hot and dry conditions, but mechanical cultivation is more effective to limit the proliferation of weed populations (Chippendale, 1979). Another problem with herbicides applied under arid conditions is that they accumulate in the soil, thereby potentially leading to serious environmental problems.

At some sites near the high-latitude and high-altitude boundaries of current agricultural production, increased temperatures can benefit crops otherwise limited by cold temperatures and short growing seasons, although the extent of soil suitable for expanded agricultural production in these regions may not be appropriate for viable commercial agriculture (Rosenzweig *et al.*, 1993). Increases in crop yields at high elevations will be the result of the positive physiological effects of CO_2, the lengthened growing season and the amelioration of low-temperature effects on growth. It can be surmised that at the upper limit of current agricultural production, increased temperatures will extend the frost-free growing season and provide regimes more favourable to crop productivity. However, in lower-latitude mountain regions, there could be causes for reductions in crop yields, in particular through a shorter growing period and a decrease in water availability. Higher temperatures during the growing season speed annual crops through their development, allowing less grain to be produced. Problems of water availability result from a combination of higher evapotranspiration rates in the warmer

Box 7.1 Climatic change in Chimborazo Province, Ecuador

In the central sierra of the Ecuadorian Andes, close to the equator, subsistence agriculture is practised at high elevations (2800–3800 m). The valley interiors of this region display a variety of regional climates. Some are drought-prone while others are moist, depending on their disposition to prevailing rain-bearing winds. In the dry valleys, rainfall generally increases with elevation, and farmers can minimize the effects of drought by cultivating crops higher up the valley sides. However, at these higher elevations, temperatures are lower and there is an increased risk of frost damage to crops. With these two dominant constraints in mind, farmers must choose which crops or crop mixes provide adequate returns in the prevailing climatic conditions while minimizing the risk of large-scale crop losses.

A detailed study has been made of the effects of climatic variations on agriculture in Chimborazo Province, a region of about 8000 km^2 in the central sierra. Here, elevation ranges from below 2000 m in the valley floors to 6675 m (the volcanic peak of Chimborazo). The upper limit of cultivation lies at about 3800 m.

Maize is the major crop at lower elevations (below 3300 m), because of its tolerance of high temperatures and drought. At higher elevations, potatoes, barley and broad beans are cultivated, along with other highland crops such as quinoa, oca, ullucu and tarwi. The high grassland (*paramo*), above the cultivation limit, is used for cattle and sheep grazing, along with lower-level land on large farms (*haciendas*) and on frost-prone flats.

Given the fall of temperature with altitude, barley requires at least 10 months to mature at 3200 m. The time required for potato maturation increases from 5–6 months at 2800 m to 9–10 months at 3800 m. Planting times tend to be earliest at the highest elevations. Sowing may occur in August at 3800 m, but as late as February below 3000 m. Multiple planting times are also employed to help limit the effects of early-season frosts.

The upper limit of intensive cultivation on flats (3080 m) is lower than on slopes (3800 m). It is probable that this limit is set by frost risk, since mean annual temperatures are higher than at the limit of cultivation on slopes. It is less clear what determines the upper limit of slope cultivation. Yields of barley and potato do not appear to diminish markedly as mean temperature declines towards the limit. However, as crop maturation time extends, it appears likely that the susceptible growth phases are pushed into unfavourable, frost-prone seasons.

Climate scenarios suggest that a decrease in late winter temperature minima of about 0.8°C could reduce the upper limit of cultivation by about 200 m. This suggests a potential loss of 8 per cent of arable land in the province. On flat land, a comparable cooling of the climate could lead to about 7 per cent of the currently cultivated area going out of production.

For a warming of the same magnitude (0.8°C), frost risk would decrease and the upper limit of slope cultivation could be raised by some 200 m to 4000 m above sea level. A rise of only 120 m would increase by over 60 per cent the area of potentially fertile flat land in the province. Under the warming scenario, it is suggested that this level of expansion could be exceeded. Moreover, evidence of prehistoric raised fields at elevations up to 3150 m in nearby Imbabura Province implies that flats have been cultivated at higher elevations in the past.

In a further analysis, the potentially cultivable area for barley and potatoes has been defined according to their respective upper and lower annual precipitation and temperature requirements. These areas were then mapped for the present-day climate and for anomalously (1-in-10 probability) dry and wet years. Under the 'dry-year' scenario (65 per cent of mean annual precipitation), total cultivable area declines by nearly 15 per cent, almost entirely owing to a contraction of potato suitability in drought-sensitive areas. In contrast, the areas that become too dry for barley are almost completely replaced by areas that would usually be too wet.

A substantial area in the interior of Chimborazo, including most of the population centres, is located in the sensitive zone for drought, according to these findings. Some of the population here have access to irrigation, or to land at higher levels. Other drought adaptation strategies are also important, such as supplemental off-farm employment, food-for-work schemes and expansion of traditional irrigation networks.

The 1-in-10 wet year (134 per cent of mean annual precipitation) produces an even greater (48 per cent) decline in total cultivable area than the dry year, with major losses occurring in both barley and potato potential. Moreover, the regions affected in wet years are quite different from the corresponding areas of impact in dry years. The vulnerable areas are located on higher-level land, often away from major roads and settlements, and difficult to reach during high-rainfall seasons. In spite of their relative remoteness, over 30 per cent of the farming population in this province would be affected by a climatic anomaly of this magnitude.

Source: adapted from Bravo *et al.* (1988) and Carter and Parry (1994)

climate, enhanced losses of soil moisture and, in some cases, decreases in precipitation.

If current climate variability remains the same, adaptive strategies such as a change in sowing dates, in genotypes and in crop rotation could counteract expected production losses. In economic terms, most agricultural model studies suggest that mid-latitude mountain regions would on average benefit from climate change, because of an overall increase in crop yields leading to lower consumer prices (IPCC, 1998). Other adaptation options include both technological advances and socio-economic options, such as land-use planning, watershed management, improved distribution infrastructure, adequate trade policy, and national agricultural programmes.

It should be noted, when considering adaptation options in the agricultural sector, that the ultimate impacts of climate on the agricultural sector may be determined in fact by non-climate factors that control the system. In Europe, for example, the main driving force is the Common Agricultural Policy (CAP) for the countries in the European Union (EU), and this also affects other countries in Europe. As a result, the consequences of climatic variations in agronomic systems that are highly regulated, such as the systems in the countries of the EU, are difficult to predict, since the crops are highly subsidized and therefore the crop prices are artificially high (IPCC, 1998). In mountain regions of Europe, particularly in Switzerland and to a lesser extent in Austria, subsidies to mountain agriculture encourage farmers to remain in these regions, thereby acting as 'caretakers' of the mountain scenery. Reductions of subsidies to such farmers, as planned following the ratification of the Uruguay Round of negotiations of the World Trade Organization, could lead to a collapse of mountain agriculture in these countries. The impacts of direct economic interference would be far more fundamental than those of expected environmental change in coming decades.

The ease of adaptation of mountain agriculture to climate change is likely to vary with crop, site and adaptation technique. Determining how countries, particularly developing countries, can and will respond to reduced yields and increased costs of food is a critical research requirement. The IPCC (1996) suggests that while global agricultural production is likely to be maintained, marginal areas such as mountains may suffer adverse consequences.

The worst-case situation for marginal agriculture would arise where climate change is severe in a country with low economic growth and capability for adaptation. In order to minimize possible production losses, food price increases and famine, it would be necessary for the agricultural sector to pursue the development of crop breeding and management programmes for heat and drought conditions; even in a highly uncertain context of climate change scenarios, such developments will in any case be useful in improving productivity in marginal environments even today.

More important in certain parts of the developing world is the potential for complete disruption of the life pattern of mountain communities which climate change may represent in terms of food production and water management. People in the more remote regions of the Himalayas or Andes have for centuries managed to strike a delicate balance with fragile mountain environments. This balance would probably be disrupted by climate change and it would take a long time for a new equilibrium to be established. In cases such as this, positive impacts of climate change (e.g. increased agricultural production and/or increased potential of water resources) are unlikely because the combined stressors, including negative effects of tourism, would overwhelm any adaptation capacity of the environment.

7.3 Human health

As with many of the socio-economic sectors in this section, it is difficult to associate any particular change in the incidence of a particular disease with a given change in a single environmental factor. It is necessary to place the environment-related health hazards in a population context.

Age, level of hygiene and socio-economic status, skin pigmentation and health status will all be determinants of the net effects of climatic change. Some examples of the associated health effects include infant mortality from diarrhoeal diseases and malnutrition (age), waterborne diseases (hygiene, socio-

economic status), risk of skin cancer (skin pigmentation) and the susceptibility of cardiovascular systems to heat. Similarly, geographical factors will determine the populations at risk from certain conditions or events and from the ensuing effects on the maintenance of traditional agriculture.

The distribution of human population over an extremely wide range of environmental temperature conditions illustrates the adaptive capacity of humans. Indeed, human beings can cope with wide extremes in environmental conditions much better than any other species, and can live in virtually every climate on earth (Weihe, 1979), in particular through technological adjustments, adapted clothing, etc. Furthermore, environmental factors can induce specific, acquired adaptations in individuals, either functional or morphological, such as the barrel-shaped thorax of people living at high altitudes, as found in miners in Bolivia, for example. Social and cultural adaptive measures, such as hygiene practices, clothing, housing, and medical and agricultural traditions, support the reversible adaptation of human beings to a particular environment. The viability of a socio-cultural adaptation is determined, among other things, by the strength of the economy, the quality and coverage of medical services, and the integrity of the environment.

Climatic changes will develop gradually over several decades. Some of them, such as heatwaves, will have a direct effect on human health (causing heat illness, for example), but many of the changes will have mainly indirect effects by changing natural ecosystems, affecting such aspects as food production, vector-borne diseases, and a number of other infectious and non-infectious diseases. These phenomena are likely to precipitate migration from one rural region to another, and from rural to urban areas. If, however, the climatic change is accompanied by an increase in intensity of some natural disasters, such as cyclones and floods, immediate effects on human health become more likely. Moreover, these catastrophes can generate large refugee and population movements, with a need for resettlement in already densely populated areas.

Changing climatic conditions exert stress on the body, requiring regulation of bodily functions to re-establish internal balance. The stress resulting from a change in climatic conditions, in turn, produces a strain within the body to establish the physiological response and to determine conscious behavioural regulation. Minor changes within daily life or diurnal climatic variations, which may necessitate adjustment or regulation, go unnoticed without amounting to stress. They do not tax the adaptive capacities of the individual. Unlike plants and non-human animals, human beings have the adaptive ability to contemplate the future and adopt appropriate strategies for protection. The human race has the freedom to alleviate or avoid a climatic stress and, in this way, to protect the body. Climatic stress can be aggravated by inappropriate behaviour or, conversely, can be ameliorated by the use of more adaptive capacities, such as awareness of the effects of temperature increases. This takes place empirically and is manifested in habits, customs and cultures that develop through a continuous interaction between human beings and their environment.

If there were a local climatic change to which an indigenous population was not accustomed, the impact of the stress would greatly depend on the ability and willingness of the population to accept the change. Even major, anthropogenic climatic changes are unlikely to be out of the range of climatic variations that have been experienced historically, and against which adaptive strategies have been developed. The total number and kind of environmental adjustments needed may be vast; they may also be expensive and may require many sacrifices in lifestyle and well-being to re-establish and maintain the basic needs. This implies that the poorer strata of the population may be the most affected and the least capable of responding to change.

7.3.1 Hygrothermal stress

Changing hygrothermal conditions can be expected to influence human health and well-being in proportion to the degree of heat stress. In healthy individuals, an efficient heat regulatory system will normally enable the human body to cope effectively with a moderate rise in ambient temperature. Thus, within certain limits of mild heat stress and physical activity, thermal comfort can be maintained by appropriate thermoregulatory behavioural responses, and physical and mental work can be pursued without detriment. Heat acclimatization will develop after several days of heat exposure, and this will help to alleviate the effects of heat stress. However, severe heat stress can result in a deterioration in health including heat illness, with effects ranging from mild, reversible cardiovascular disturbances to severe tissue damage and death. Of primary concern are certain high-risk groups in whom even mild heat stress may produce abnormal heat strain and heat-related disorders. Global warming in some regions may be accompanied by an increase in relative humidity, which, in warm environments, greatly increases heat stress and thermal discomfort, because of reduced evaporative cooling from the skin surface. Although hygrothermal conditions can be expected to exert thermal stress on populations, adaptation will readily occur with prolonged and gradual warming. Extreme variations and rapid changes in thermal conditions, especially in low latitudes with existing high heat stress, and in densely populated urban areas, will carry an increased risk of heat-related disorders. Mountain regions will often be seen as a pleasant alternative to the hot lowlands, and may witness a large influx of persons seeking cooler conditions during the warmest part of the year.

7.3.2 Effects of ozone depletion

Current recorded measurements indicate that there has been a significant decline in stratospheric ozone concentrations over the northern hemisphere of 1.7–3 per cent during the past 20 years. While the ozone decrease has been small, but significant, the increase in biologically effective ultraviolet radiation (UV-B) exposure has to date been minimal. This is due to the seasonal distribution of ozone reduction. The major effects occur in the winter months when, because of the low sun angle and therefore a long path through the atmosphere, UV-B radiation at ground level is normally low. There are also significant seasonal variations. Thus, the increases in UV radiation at the Earth's surface over ambient levels are quantitatively very small (Dahlback, 1989). Estimates have been made that, for a 1 per cent decrease in stratospheric ozone, UV-B radiation could increase by 1.25–1.5 per cent (Urbach, 1989).

The changing ozone column has three different kinds of effects: one biological, due to increases in exposure to UV-B radiation; one chemical; and one related to climatic change. The health effects, occurring largely as a result of increases in biologically effective UV radiation, are expected to consist of increases in certain forms of skin cancer and eye diseases (primarily cataract) and possible alterations in human immune response.

In certain high-altitude mountain regions, particularly in Europe and North America, there may be a small but significant increase in UV-related health problems because of the fact that large numbers of persons practise winter sports and, without adequate protection, risk greater exposure.

7.3.3 Air pollution

It is probable that atmospheric loadings of pollutants such as NO_x, tropospheric ozone, pesticides, volatile organic compounds, and soot particles and other aerosols may be influenced by global warming, but it is not possible to predict with accuracy how the frequency distributions will change at a specific location. Changes in the frequency and duration of atmospheric stagnation are likely to take place, and, as a result, modifications may be needed to emission-control implementation plans that have been designed on the basis of past air pollution and meteorological statistics. There is little doubt that air pollution causes increased respiratory morbidity and mortality (Whittemore, 1985). Ozone, in concentrations measured in ambient air, has been shown to irritate and cause inflammatory reactions in the respiratory tract of sensitive individuals

(Schneider, 1989). However, in combination with other pollutants, such as acids, the compounded effects are often much greater. A number of different pollutants can interact chemically in the atmosphere and produce more toxic products. Both ozone and peroxy-acetyl nitrate (PAN) are phytotoxic and their effects on food crops and plants are also of indirect importance to human health.

Because of their relatively sparse population and reduced level of industrial activity, mountain regions are not in themselves major source regions for most anthropogenic pollutants. However, because of the potential for long-range transport within the atmosphere, mountain populations are certainly at risk from pollutants originating far afield.

Apart from the effects of global climatic change, air pollution is expected to increase throughout the world because of extending industrialization. The precise impact of global warming on air pollution and population exposures cannot be clearly estimated. However, the potential exists for major changes in the concentrations, duration and types of pollution, all of which could significantly affect air pollution-related morbidity and mortality in populations, not only in the source regions themselves but in many mountain regions located in the distant periphery of the industrial areas.

7.3.4 Vector-borne and non-vector-borne diseases

When discussing the impact of the projected climatic changes on communicable diseases, two basic mechanisms can be distinguished. The first mechanism operates through changes in the ecology of vectors, mostly arthropods, of a series of diseases that are currently prevalent, mostly in tropical and subtropical regions. The second mechanism is through a direct modification of human-related risk factors, such as the availability and quality of water for domestic and agricultural purposes. The occurrence of vector-borne diseases is widespread, ranging from the tropics and subtropics to the temperate climate zones. With few exceptions, they do not occur in the cold climates of the world, and are absent above certain altitudes in mountain regions of the tropical and equatorial belt.

The occurrence of vector-borne diseases is largely determined by the abundance of vectors and intermediate and reservoir hosts; the prevalence of disease-causing parasites and pathogens suitably adapted to the vectors, the human or animal host, and local environmental conditions, especially temperature and humidity; and the resilience and behaviour of the human population. Climate change will impact upon each of these factors.

Vectors require specific ecosystems for survival and reproduction. These ecosystems are influenced by a number of factors, many of which are climatically controlled. Changes in any of these factors will affect the survival and hence the distribution of vectors (Kay *et al.*, 1989). The global climatic change projected by the IPCC (1996) may have a considerable impact on the distribution of vector-borne diseases. A permanent change in one of the abiotic factors may lead to an alteration in the equilibrium of the ecosystem, resulting in the creation of either more or less favourable vector habitats. At the present limits of vector distribution, the projected increase in average temperature is likely to create more favourable conditions, in terms of both latitude and altitude for the vectors, which may then breed in larger numbers and invade formerly inhospitable areas. Increases in temperature coupled with regional increases in rainfall may lead to an expansion of habitats for malaria vectors, whereas reduced rainfall associated with such temperature increases may create new habitats for the phlebotomine vectors of leishmaniasis.

The anticipated shifts in temperature and precipitation in a warmer global climate will affect vector reproduction and longevity. Hence, disease transmission that is at present seasonal may change to perennial transmission and vice versa. Regions at higher altitudes or latitudes may become hospitable to the vectors. Thus, disease-free highlands, such as parts of Ethiopia, Indonesia and Kenya, may be invaded by vectors as a result of an increase in the annual temperature.

The World Health Organization (WHO, 1990) has investigated the possible changes in the distribution of the principal vector-borne diseases. For malaria, which is a major cause of mortality in certain developing countries, rises in temperature and rainfall would most probably allow malaria vectors to survive in areas immediately surrounding their current distribution limits. How far these areas would extend latitudinally and altitudinally would depend upon the extent of warming. In certain countries where the disease has been eradicated in the second half of the twentieth century, particular strains of malaria are resurging. There are reports from various low- to medium-elevation upland sites in Turkey, Tajikistan, Uzbekistan, Turkmenistan and the Urals that malaria is being transmitted in rural populations (John *et al.*, 1998), with close to epidemic proportions in south-eastern Anatolia.

The distribution patterns of another tropical disease, schistosomiasis, are determined by the presence of suitable intermediate hosts, and by activities involving contact with fresh water, including labour-intensive agricultural practices, which contribute to increased transmission. An increase in the frequency of climatic conditions favourable for the development of the disease may lead to a significant expansion of risk areas, provided that the intermediate hosts invade previously unfavourable areas and are able to compete successfully for survival. In the case of African trypanosomiasis ('sleeping sickness'), serious epidemics occur at times when all the ecological elements necessary for effective transmission are favourable in some areas, notably East Africa. Higher rainfall in the forest belt may lead to increased breeding of the tsetse fly, the only known vector of human African trypanosomiasis. Climatic change could also affect the development of the parasite in the vector.

As is the case for vector-borne diseases, most of the non-vector-borne diseases are concentrated in populations living in tropical and subtropical regions of the world. While for vector-borne diseases, water plays a role as the habitat of the vector (Feacham *et al.*, 1980) or as a necessary factor for its propagation, in

Box 7.2 Threats from malaria and other diseases in tropical mountain regions

On Indonesia's Carstensz Mountain in Irian Jaya (the Indonesian part of New Guinea), ice always covers the mountain top. Also known as Puncak Jaya Kesum, it is the only mountain in the Asian tropics which year after year retains an icecap. But the mountain's glaciers have receded dramatically over the past few centuries, owing to at least in part a warming of roughly 0.5°C. As a result, the snow line has risen about 100 m up the mountains. This sign of global change has triggered other symptoms of long-term warming which are likely to continue into the future, namely the upward migration of certain plant species and the opening up of new regions favourable to diseases confined to lower elevations.

As warmer conditions push up the mountain, higher-altitude environments become more favourable for certain organisms, including agents of disease such as malarial parasites, and disease vectors such as mosquitoes. The *Aedes aegypti* mosquitoes, for example, are known to spread dengue and yellow fever. Previously limited around the world to altitudes no higher than 1000 m, they are now reportedly found in Colombia at elevations above 2200 m. Epstein (1992) cites similar types of upward migrations in Central America and in East Africa.

Loevinsohn (1994) has reported that malaria has in recent years established itself in high-altitude zones of Rwanda, where it had previously been absent or at least rare. He attributes the spread of malaria in part to a large increase in mean temperatures over the past three decades. Malaria epidemics have also occurred in other parts of East Africa since the late 1980s, and Loevinsohn suggests that these too were driven in part by heavy rains and warmer temperatures, which could become the norm in coming decades.

Martens *et al.* (1995) suggest in their study that the range of plausible climate scenarios put forward by the IPCC (1996) would result in a widespread increase in risk due to expansion of the areas to higher elevations suitable for malaria transmission. Martin and Lefebvre (1995) have expressed concern over the spread of the disease to areas where people have acquired little or no immunity, which is generally the case in mountain regions. In their study, they note that the spread of seasonal malaria is most likely to foster epidemics, causing widespread debilitation and increased mortality.

---

the case of waterborne diseases it can also act as the habitat and vehicle of the pathogen itself. In this case, the infection in humans occurs through the direct consumption of the water or through contamination of food. Furthermore, a lack of a sufficient quantity of household water invariably leads through poor personal and food hygiene to other forms of water-propagated diseases. These types of disease are largely conditioned by poverty, linked to poor sanitary conditions. Among these diseases, childhood diarrhoea is probably the one taking the heaviest human toll, in terms of years of potential life lost.

Populations in remote mountain regions of tropical Africa and Asia are particularly vulnerable to such diseases.

7.3.5 Possible adaptation options

The World Health Organization (WHO, 1990) urges national health authorities to install or reinforce surveillance systems for the health problems that are most likely to be influenced by climatic change. Such action would be aimed at facilitating the gathering of baseline data for the evaluation of time trends or outbreaks, the definition and evaluation of priority fields for action or research, and the evaluation of actions taken.

Since global warming might lead to a redistribution of vector-borne diseases, national and international efforts would be required for their control. In many instances where these diseases are already prevalent, such efforts are inadequate or lacking. The national health authorities should be adequately prepared to handle the control of vector-borne diseases through vector control, the vaccination of individuals at risk, and drug treatment. In view of a possible increase in the incidence of malnutrition arising from a disruption in food supplies, the adoption of new crops or new agricultural production techniques should be fostered, particularly for the areas that might be affected.

It is important to increase awareness of the potential health effects of changes in the climate. For this purpose, region-specific educational material that addresses particular local health problems should be produced. Such materials should be distributed to professionals (physicians, epidemiologists, architects, etc.), schools and the general population.

7.4 Tourism

Tourism is of great economic importance and is one of the fastest-growing economic sectors in the world. It accounts for 10 per cent of the world's real net financial output, but many countries and in particular those in the developing world are dependent on tourism to a far greater degree than in the industrialized world. In the developing countries, tourism of all types contributes roughly US$50 billion annually (Perry, forthcoming). Even in the current period of widespread economic recession and depression, tourism has remained surprisingly strong. Figures released by the World Tourism Organisation indicate that the number of international tourists has increased 25-fold in the second half of the twentieth century. It is estimated that if current trends continue, international tourism will double every 20 years. Furthermore, tourism patterns have become more diversified: new activities

have joined traditional recreational patterns. As a consequence, even remote and so far untouched natural areas, in particular mountain regions in the Himalayas, the Andes and East Africa, are being visited by tourists more frequently.

The major trends in international tourism are characterized by higher demands for air travel to remote destinations, and an increasing trend towards various forms of tourism in natural areas, such as climbing, kayaking, diving, hang-gliding or snow-boarding. These activities often take place in highly sensitive mountain regions which warrant particular protection. In order to keep up with the increasing number of tourists in mountain regions of the world, there is a parallel boom in the development of tourism infrastructure and construction in attractive cultural and natural landscapes, which can be detrimental to those landscapes and the sensitive ecosystems which they support.

Climate change is likely to have both direct and indirect impacts on tourism in mountain areas. Direct changes refer to changes in the atmospheric resources necessary for specific activities. Indirect changes may result from both changes in mountain landscapes, which Krippendorf (1984) refers to as the 'capital of nature', and wider-scale socio-economic changes such as patterns of demand for specific activities or destinations and for fuel prices.

The marked seasonality of mountain climates implies that their attractions for tourists vary greatly through the year. Various methodologies have been developed to assess the suitability of regions for specific activities in different seasons. While such approaches are based on long-term averages, others have been developed to assess the economic implications of historical climate variability over short periods, mainly for the skiing industry (Perry, 1971; Lynch *et al.*, 1981).

In the European Alps and the North American Rockies in particular, the ski industry is for some resorts by far the greatest single source of income. In many mountain communities, there is no alternative to skiing capable of generating such major financial

Box 7.3   The importance of tourism in Nepal

Nepal is a country rich in natural and cultural diversity, which features some of the highest mountain chains in the world. However, the poverty and the rapidly deteriorating environment urgently requires new approaches to developing sustainable livelihood systems that make the most of the comparative advantages that mountains have over the plains. Mountain tourism in these regions is often an alternative to subsistence agriculture, and can generate employment opportunities and improve individual and community income.

Tourism began in Nepal in the late 1960s, at a time when the country was relatively inaccessible, and was constrained by the lack of international air transport services, as well as of other facilities and services required by tourists. However, tourist arrivals in the country increased with an average growth of over 6 per cent in the last 20 years of the twentieth century. The gross earnings from tourism increased from 1980 to 1994 at an average annual growth rate of about 17 per cent. The share of earnings from tourism in the total value of merchandise export has fluctuated between 35 and 55 per cent, and its share in total foreign exchange earnings has remained fairly constant. The average contribution of tourism earnings to GDP increased from 2.3 per cent in 1980 to about 4.2 per cent in 1994. However, the benefits of tourism have been confined to very localized areas, and only a small number of people have been able to take advantage of this new industry. Development has not been able to reach many remote and inaccessible areas of the mountains, and as a consequence many indigenous communities still depend on natural and land resources for subsistence. With the population expanding rapidly, the environment is suffering significant degradation.

Given its relatively small size, the biodiversity of Nepal is immense. In order to protect this biodiversity, 14 protected areas covering roughly 14 per cent of the country have been created. Trekking tourism is the most popular type of mountain tourism in Nepal, followed by mountaineering and rafting.

A number of studies suggest that protection has enhanced conservation. But since most of these areas were inhabited by people before the protected areas and national parks were created in Nepal, it has also compelled people to bring changes to their traditional lifestyles. Conflicts between park authorities and local communities have also emerged, as development and tourism have not been able to benefit the people that reside in protected areas.

Serious concerns have been raised on the negative impacts of mountain tourism on the environment, although not all of these can be attributed solely to tourism. Negative environmental impacts often impose more hardship to the mountain people who depend on natural resources which provide the basis for mountain tourism in Nepal. Thus the deterioration of the resources which the natural environment provides is likely to have increasingly serious negative implications for mountain tourism itself.

Mountain tourism is popular in many of the national parks, such as the Annapurna Conservation Area, Sagarmatha National Park and Langtang National Park. Over 50 per cent of the trekkers who come to Nepal visit the Annapurna Region, which is easily accessible and located about 200 km west of Kathmandu. The Sagarmatha National Park contains the world's highest mountain, Mt Everest (8884 m), a legendary summit which attracts numerous trekkers. Many of the other mountain national parks which are not so readily accessible exert less attraction for tourists.

Source: adapted from Chalise (1997)

resources. Capital investment for cable-cars, ski-lifts and chairlifts in countries such as Austria, Switzerland, France and Italy needs 20–30 years for a positive return on investment. Lower revenues would put these investments at risk, which would impact negatively on the financial revenue of mountain ski resorts, which are often the major shareholders in the cable-car and ski-lift companies. In an attempt to alleviate problems related to the lack of snow at lower elevations, recent investments such as in snow-making equipment have been made in Europe and the United States. However, in a warmer climate as projected for the twenty-first century, these snow-making machines would be of little use as temperatures need to be well below freezing for them to be effective. Given the prospects of aperiodic snow, and deficient winters in many mountain resorts, it is likely that tourists will either no longer come during the peak seasons, or may delay booking accommodation until snow is secure. In both cases, this will lead to reduced sales and rental of ski gear, and a consequent increase in partial or full unemployment in sectors dependent on skiing, namely hotels, restaurants, sports stores, etc.

Using scenarios derived from GCMs, a number of investigations have been carried out to examine the possible implications of climate change for skiing in Australia, eastern Canada and Switzerland (McBoyle and Wall, 1987; Galloway, 1988; Lamothe and Périard, 1988; Abegg and Froesch, 1994). Abegg and Froesch (1994) have shown in their study of the Swiss ski industry that if temperatures were to rise by about 2–3°C by the year 2050, the low- to medium-elevation resorts located below 1200–1500 m above sea level would be adversely affected. Warmer winters bring less snow, and the probability of snow lying on the ground at peak vacation periods (Christmas, February and Easter) would decline. A general rule for the viability of the ski season in Europe is a continuous snow cover of over 30 cm depth for at least 100 days. On the basis of these figures, Abegg and Froesch (1994) have shown that while towards the latter part of the twentieth century, 85 per

cent of ski resorts have reliable amounts of snow for skiing, a 2°C warming would bring this figure down to 63 per cent. Regions such as the Jura Mountains to the west of the country, whose average altitude lies between 900 and 1200 m, would seldom experience significant periods of snow cover, whereas the elevated ski resorts in the central and southern Alps would be less severely affected.

The economic impacts on ski resorts of changing patterns of snowfall in a changing climate may appear to be far removed from the preoccupations of communities in the mountains of certain developing countries. Yet ski resorts are also found in the Andes (Fuentes and Castro, 1982; Solbrig, 1984) and the Himalayas, and changes in the length of the snow-free season would be of critical importance for most mountain communities. In South Asia, another important potential change for communities which increasingly depend on tourism concerns the monsoon, whose timing may well change (IPCC, 1996). This could have substantial effects on countries, such as Nepal and Bhutan, for which tourists are the principal source of foreign exchange (Richter, 1989).

To some extent, such impacts might be offset by new opportunities in the summer season and also by investment in new technology, such as snow-making equipment, as long as climatic conditions remain within appropriate bounds. Mountaineering and hiking may provide compensation for reduced skiing, and thus certain mountain regions would remain attractive destinations. However, global climate change has wider implications for traditional holiday breaks, with destinations other than mountains in winter becoming far more competitive. Higher temperatures imply longer summer seasons in mid-latitude countries, and Perry and Smith (1996) have noted that these may well result in a new range of outdoor activities. In the Mediterranean region, much warmer temperatures will have negative health impacts related to heat stress and skin cancers, and mountains in the vicinity of many Mediterranean beaches are likely to offer a cooler alternative to the hot beaches. Mountainous islands such as Corsica,

Sardinia or Cyprus, or the coastal ranges in Spain, Italy and Greece, may continue attracting tourists who traditionally spend their vacation at the seashore.

In addition to these potential direct impacts of climate change on tourism, a critical indirect impact needs to be emphasized. One of the most likely types of policy response to climate change will be the imposition of 'carbon taxes' on fossil fuels (Bryner, 1991). These will increase the costs of fuels, a major component of the cost of tourism, and in particular to mountain regions, which are generally not readily accessible. Other indirect impacts might include decreasing attractiveness of landscapes, and new competition from other tourist locations as climate changes.

Today, very few communities can rely on tourism as a reliable source of year-round employment and income. This is a particular problem in the many ski resorts where off-season work is limited because they are not based on pre-existing settlements. Seasonality is important not only in terms of employment, but also because facilities built for tourism represent investments that must be maintained year-round and paid off (Barker, 1982).

In both developing and industrialized countries, tourism has led to considerable changes in patterns of agricultural production, forestry and water resources management, and employment (Price, 1994). Thus, the implications of climate change for tourism must be regarded in conjunction with the implications for other resources (Breiling and Charamza, 1994).

Towards the latter part of the twentieth century, travel with nature as a principal objective has become a major segment of the tourist industry. Known under a variety of names, nature-based tourism is promoted by the travel industry as a unique opportunity to see and experience natural environments and local customs in ways not available to participants in mass tourism. Nature tourism is developing a popular following, and it is seen by some as one of the central elements in sustainable economic development. This tendency is likely to expand rapidly in coming decades.

An unspoilt natural environment is essential for successful tourism in mountain regions. Moreover, many tourists are seeking an intact natural environment and, increasingly, spectacular natural scenery. In addition to traditional tourist destinations in the mountains, more and more unusual and particularly attractive natural areas have been developed for tourism, such as high mountains and glaciers, waterfalls and natural watercourses. In many regions, remote and barely accessible zones are being opened up to adventure-loving tourists. This is increasingly leading to a conflict of interests in terms of economic development versus environmental conservation, because many of these areas are ecologically sensitive and have been designated as protected areas for precisely this reason. Current trends in international and domestic tourism raise a number of problems for biological diversity. The increasing number of visitors as well as the intensity of tourist activities may have both direct and indirect negative impacts on the natural environment, particularly in ecologically sensitive areas. Expanding tourism developments and infrastructure use up the resources of the natural environment and the countryside. They also lead to environmental stresses, such as air pollution, noise and the consumption of resources, and may deflect budgets away from resource management.

'Ecotourism' is today unfortunately used as an all-inclusive term; since the 1980s, it has become increasingly popular to be 'environmentally conscious', and some individuals and organizations have begun to advertise their products and services as environmentally friendly in such a manner that nearly all travel qualifies. This adversely affects protected areas and biodiversity in several ways. When correctly managed, however, ecotourism can be perceived as being in the best interests of both local communities and the travel industry. Ecotourism needs to have a range of constraints and quality labels with precise requirements. It is especially important that the requirements be respected, and that the public be informed and motivated to insist on compliance.

Ecotourism is often defined as sustainable nature-based tourism. However, ecotourism also incorporates social and cultural dimensions, where visitors interact with local residents in national parks, remote areas or indigenous homelands. Indigenous peoples are frequently the main attraction for ecotours visiting wild and scenic natural areas in remote mountains of South America, Asia, Oceania and Africa. Native lands in developed countries are also a growing focus for indigenous ecotourism. To benefit local communities and be socially sustainable, ecotourism needs to foster environmental and cultural understanding, appreciation and conservation.

When ecotourism projects are developed in consultation with host communities there are several mutual benefits. Tour operators gain access to remote areas and local mountain communities. Local people derive income from hosting visitors while elders pass on cultural knowledge. The benefits of ecotourism for rural or indigenous communities include preservation of cultural traditions, conservation of the natural environment and maintenance of social, cultural and religious values. In remote areas with limited development, ecotourism ventures can improve the quality of life and well-being of local and indigenous communities. However, there is concern that in many instances there is an insufficient degree of participation by the local population in the planning and implementation of tourist activities, and in the distribution of the profits of tourism. There is also concern that tourism can have negative cultural impacts which may be detrimental in the short term to the world's cultural heritage in many regions.

Ideally, tourism should have a positive rather than a negative impact on biological diversity. In order to achieve this, development of tourism needs to be managed. Tourism needs to be planned at the appropriate level, whereby tourism planning must be integrated into general development schemes. The local population need to be involved in the planning and it must be ensured that local people benefit from tourist activities.

The management of tourism also implies that tourist activities may need to be regulated, particularly in protected and ecologically sensitive areas, even where those areas do not have the formal status of protected areas. Tourist activities need to be assessed to ensure that they comply with the requirements of the conservation and sustainable use of biological diversity. It is also necessary to establish control mechanisms to enforce the decisions concerning environmentally friendly tourist activities. Tourism is a continuously adapting industry which responds to changing demographic, economic and technological conditions, as well as to new demands from the tourists themselves. In view of the fragmented nature of the tourist industry, its adaptation to climate change will probably be gradual, and new investments will take place in parallel to other strategic decisions (Perry, forthcoming).

Tourism generates the need for more information about other lands, their geographical characteristics and their people. It can draw the attention of people and governments to values of culture and nature, to necessities for conservation and education programmes related to them. The discovery of other peoples' land and heritage in a respectful way can create a network of positive relations to nature, culture and people. Tourism should be a self-sustaining activity, respectful of the environmental, social and economic carrying capacities. Tourists are only guests in the host country and should therefore behave appropriately. Planning processes are a fundamental means of achieving sustainable tourism. The ill-conceived and badly planned development of tourism will inevitably harm or even destroy nature, monuments and indigenous human societies and cultures. These planning processes must recognize the needs and aspirations of the various component communities, thereby requiring an inventory of natural and cultural resources of the region. They should determine the biophysical, economic, environmental and other constraints to growth and to development, establish land-use priorities and identity areas for development and for conservation, determine the limits of acceptable change to the area in a tourism context, and integrate tourism with other land uses.

7.5 Hydro-power and other commercial activities

An important socio-economic consequence of global warming on the hydrological cycle is linked to potential changes in runoff extremes. Not only the mountain population but also the people in the plains downstream (a large proportion of the world population) presently depend on unregulated river systems and thus are particularly vulnerable to climate-driven hydrological change. Current difficulties in implementing water-resource development projects will be compounded by uncertainties related to hydrological responses to possible climatic change. Among these, possible increases in sediment loading would perturb the functioning of power generating infrastructure.

Thermal, nuclear and hydro-power stations rely on the supply of water for cooling or for the direct generation of electricity. Many of the more important hydro-power dams are located in mountains, where the head of water can reach considerable heights in Switzerland, Austria, Norway, Russia, the United States and New Zealand. Changes in flow regimes, induced by changes in either total precipitation, the amount of snow-melt, or a combination of both, would affect hydro-power potential. Citing pan-European research examples, Arnell (1999) mentions that a shift in peak discharge rates from spring to winter in Norway would reduce power-generation potential in spring, but would increase it in winter during the peak demand season. This would not necessarily be the case in other regions, such as Greece, where the reduction of spring and autumn flows could shift the seasonality of power generation and lead to reductions of 5–25 per cent by the 2050s. There is a technical problem here, since electricity cannot be stored in any significant amount; it is thus critical that it be generated precisely when required.

The sensitivity of some thermal and nuclear power generating stations to shortfalls in water for cooling purposes may increase in the future, particularly during the summer months. Lack of water can lead to reduced or intermittent energy production, for obvious security reasons. There may be a real risk of increases in such reductions in energy production in coming decades, particularly in those areas which are likely to become drier in the future.

Sensitivity of mountain hydrology to climate change is a key factor that needs to be considered when planning hydro-power infrastructure. In the future, a warmer and perhaps wetter greenhouse climate needs to be considered. The impact of climate on water resources in alpine areas has been examined by Gleick (1986, 1987a,b) and Martinec and Rango (1989). Similar studies have related electricity demand to climate (Warren and LeDuc, 1981; Maunder, 1986; Downton et al., 1988). However, few have attempted to integrate these impacts of climate change by considering both electricity supply and electricity consumption (Jaeger and Kellogg, 1983).

Mountain runoff (electricity supply) and electricity consumption (demand) are both sensitive to changes in precipitation and temperature. Long-term changes in future climate will have a significant impact on the seasonal distribution of snow storage, runoff from hydroelectric catchments and aggregated electricity consumption. On the basis of a study made in the Southern Alps of New Zealand, Garr and Fitzharris (1994) have concluded that according to future climate scenarios used (New Zealand Ministry for the Environment, 1990), the seasonal variation of electricity consumption will be less pronounced than at present, with the largest changes in winter, which corresponds to the time of peak heating requirements. There will also be less seasonal variation in runoff and more opportunity to generate power from existing hydro-power stations. The electricity system will be less vulnerable to climate variability in that water supply will increase, but demand will be reduced. These conclusions suggest that climate change will have important implications for hydroelectricity systems in other mountain areas as well.

The countries of the Hindu Kush–Himalaya region are currently undergoing rapid economic transition in order to meet their overall requirements for development purposes. This includes energy demand, which until recently depended almost entirely upon on fuel-wood, which has been a critical factor for deforestation in the region. However, forests are no longer considered to be the only source of energy, and water is now a principal source for economic development (Verghese and Ramaswamy, 1993).

The obstacles to harnessing hydro-power in countries such as Nepal or Bhutan are of an economic, technical and political nature. So far, poverty appears to be the main limiting factor for developing hydro-power potential (Chalise, 1997). The complexity of the problems associated with the development of water resources in the region could be further complicated by the potential impacts of enhanced monsoon rainfall and intensity due to global warming. The implications of such an increase in precipitation amounts in the geologically active high mountain environments of the Hindu Kush–Himalaya may be quite significant, and increased sediment loading could severely damage turbines and dam infrastructure, leading to prohibitive maintenance costs for the countries concerned.

Commercial utilization of mountain forests can be affected directly and indirectly by climate change. Direct effects include loss of viability of commercial species, including problems in regeneration and lower seedling survival. Indirect effects relate to disturbances such as fire, insect and disease losses. These indirect effects depend on the influence of climate on the disturbance agents themselves.

Many of the commercially viable mineral deposits in the world are located in mountain regions. While climate has only a minor direct influence on exploitation of these resources, it may exert a significant indirect influence. Mining causes surface disruption and requires roads and other infrastructure. Changes in climate that lead to increases in precipitation frequency and/or intensity may exacerbate the potential for mass wasting and erosion associated with these developments. Furthermore, the economics of mineral exploitation often requires processing of the extracted ore *in situ*, for example smelting and hydrochemical processing. In the latter case, climate, especially precipitation and temperature, is a critical factor in process design.

Conclusions:
adaptation strategies and policy

8.0 Chapter summary

This final chapter enters into the realm of possible policy strategies for reducing, or at least adapting to, the consequences of environmental change. A number of international agreements exist today which go some way towards providing a legal and economic framework to tackle major environmental issues. While much of the discussion contained in this chapter is not solely aimed at mountain regions, a number of principles within the charters of the conventions signed at the United Nations Conference on Environment and Development (Rio de Janeiro, Brazil, June 1992) have relevance when implemented appropriately. An adequate set of policy measures and a reasonable balance between economic interests and environmental sustainability could help reduce the risk of major environmental change in mountains and uplands.

8.1 Introduction

Throughout this book, it has been seen that mountains and uplands represent unique, sometimes marginal, elements of the Earth system. Their value in terms of biodiversity, water and other resources is unquestioned; they are often spectacular regions which attract large numbers of visitors. Many existing mountain protected areas are empty of resident

people, or nearly so; but there are others which have substantial populations, especially in the valleys. Here the landscapes have often been shaped by long occupation, and many of their special qualities lie in the contrast between the managed and natural environments. These people, over centuries, have reached a way of life which is broadly in balance with the setting in which they live. Now the fragile equilibrium which in many remote areas existed between indigenous populations and their environment is being disturbed; in certain industrialized countries, this disturbance occurred in the nineteenth and early twentieth centuries.

It is only towards the latter part of the twentieth century that environmental awareness has risen, and the need for some form of protection for mountain areas is increasingly under discussion. The degree of environmental concern varies considerably in relation to a large number of interacting social and economic factors; it is important to emphasize that environmental change may not only have negative impacts but also generate new opportunities (Glantz *et al.*, 1990). Because of the high diversity of mountains, communities in mountain regions of the industrialized countries, whose lifestyles are based on the opportunities and constraints of diverse environments, may be able to respond to environmental changes with relative ease, unless rates of change are too abrupt. However, mere knowledge of these complex environments will not be adequate; suitable societal structures must be in place or will need to be developed. The extensive networks of community and mutual cooperation that characterize many mountain communities (Beaver and Purrington, 1984; Guillet, 1983; Viazzo, 1989) may already be under stress as populations grow and change in structure, and increasing numbers of tourists arrive. However, such networks are likely to be increasingly essential, particularly in communities that are geographically or culturally distant from centres of power.

Mountain environments merit special consideration in development policy. The latter part of the twentieth century has witnessed a steady increase in global attention to mountain regions, and mountains have emerged in the sustainable development arena, notably through the mountain development chapter of Agenda 21 (see Box 8.1) of the 1992 UN Conference on Environment and Development (Stone, 1992). Despite this rising awareness, however, and the fact that sustainable and equitable development has been added to the policies of many organizations, there has been a concurrent loss of revenue and control by many mountain populations. Mountain ecosystems and biodiversity have deteriorated and the resources upon which these populations depend have dwindled. The emergence of new mountain countries, particularly in Central Asia and the Caucasus, is an important phenomenon of the 1990s that has not yet been addressed in terms of mountain development.

However, the situation related to environmental and development issues is perhaps even more complex than just described. In certain regions such as the European Alps, economic forces have changed the way of life of mountain populations in such a way that over 50 per cent of the alpine population now live in urban areas; in many mountain valleys, industry, transportation and service industries have experienced considerable development and have taken over from traditional mountain agriculture as the principal source of income. An additional consequence is that, by reducing the human demand on agricultural land, the natural forest cover is returning to its original surfaces. However, this evolution could be sharply disturbed by decisions taken far away from the Alps: decisions taken in Brussels, seat of decision-making within the European Union, could lead to economic restructuring in the primary, secondary and tertiary sectors to a degree which may be difficult to predict and therefore to plan upon well in advance. In the short term (decadal time scales), these political and economic factors are likely to lead to impacts for mountain economies, land use and landscape of a far more significant nature than climatic change or other environmental stresses (Messerli, personal communication, 1999). In the developing world, particularly in the tropics, however, mountain regions are often far more attractive for the primary sector than the neighbouring regions, which are generally too warm or too moist for exportable agricultural products. Mountain agriculture and its development, in response to rapidly growing populations in many tropical countries, is currently leading to an overexploitation of land and soil resources. This in turn increasingly results in erosion, deforestation and general environmental degradation, a situation which is difficult to reverse through political decision-making.

Because most mountain protected areas are generally chosen for their physical, biological and scenic qualities, the role of local people has usually been seen as one of maintaining these attributes. Their own way of life, the protection of it from unnecessary disruption, and the conservation of the resources upon which it is based have not been seen as features requiring protection and upon which areas should, as a consequence, be selected for protection.

The selection of protected areas in the mountains needs to be related primarily to the sets of values which it is desirable to protect. These include physical features, biodiversity, catchment characteristics, human cultures, resources and scenery. Many of the difficulties of managing mountain protected areas are concerned with establishing the correct balance between protection and use. Decentralization of planning and control

Box 8.1   Research recommendations from Agenda 21, Chapter 13: sustainable mountain development

Two research areas have been identified as requiring special attention in the context of the mountain regions chapter of Agenda 21, namely:

A. Generating and strengthening knowledge about the ecology and sustainable development of mountain ecosystems,
B. Promoting integrated watershed development and alternative livelihood opportunities.

The recommendations under research area A comprise the following elements:

(a) to undertake a survey of the different forms of soils, forest, water use, crop, plant and animal resources of mountain ecosystems, taking into account the work of existing international and regional organizations;
(b) to maintain and generate database and information systems to facilitate the integrated management and environmental assessment of mountain ecosystems, taking into account the work of existing international and regional organizations;
(c) to improve and build the existing land/water ecological knowledge base regarding technologies and agricultural and conservation practices in the mountain regions of the world, with the participation of local communities;
(d) to create and strengthen the communications network and information clearing-house for existing organizations concerned with mountain issues;
(e) to improve coordination of regional efforts to protect fragile mountain ecosystems through the consideration of appropriate mechanisms, including regional legal and other instruments;
(f) to generate information to establish databases and information systems to facilitate an evaluation of environmental risks and natural disasters in mountain ecosystems.

Nearly half of the world's population is affected in various ways by mountain ecology and the degradation of watershed areas. About 10 per cent of the Earth's population live in mountain areas with higher slopes, while about 40 per cent occupy the adjacent medium- and lower-watershed areas. There are serious problems of ecological deterioration in these watershed areas. For example, in the hillside areas of the Andean countries of South America a large portion of the farming population is now faced with a rapid deterioration of land resources. Similarly, the mountain and upland areas of the Himalayas, South-East Asia and East and Central Africa, which make vital contributions to agricultural production, are threatened by cultivation of marginal lands owing to expanding population. In many areas, this is accompanied by excessive livestock grazing, deforestation and loss of biomass cover.

Soil erosion can have a devastating impact on the vast numbers of rural people who depend on rain-fed agriculture in the mountain and hillside areas. Poverty, unemployment, poor health and bad sanitation are widespread. Promoting integrated watershed development programmes through effective participation of local people is a key to preventing further ecological imbalance. An integrated approach is needed for conserving, upgrading and using the natural resource base of land, water, plant, animal and human resources. In addition, promoting alternative livelihood opportunities, particularly through development of employment schemes that increase the productive base, will have a significant role in improving the standard of living among the large rural population living in mountain ecosystems.

The objectives of research area B include the following:

(a) by the year 2000, to develop appropriate land-use planning and management for both arable and non-arable land in mountain-fed watershed areas to prevent soil erosion, increase biomass production and maintain the ecological balance;

(b) to promote income-generating activities, such as sustainable tourism, fisheries and environmentally sound mining, and to improve infrastructure and social services, in particular to protect the livelihoods of local communities and indigenous people;

(c) to develop technical and institutional arrangements for affected countries to mitigate the effects of natural disasters through hazard-prevention measures, risk zoning, early warning systems, evacuation plans and emergency supplies.

processes within agencies and governments can contribute substantially to effective conservation and sustainable development in mountain regions.

A commitment by national governments to decentralize is critical at all levels, and across all sectors. Political liberalization should be accompanied by similar progress in the economic and social service sectors. As efforts are made to delegate decision-making and resource control, accountability and responsibility must be strengthened at the most local level. National economic and political interests need to be balanced with basic needs of mountain populations. The environmental and social impacts of industrial activities, and equitable terms of trade for mountain peoples and products, need to be carefully assessed.

National governments may benefit from a re-evaluation of biological and cultural resources in their mountain regions, particularly in the context of sustainable and equitable development. New technologies such as geographic information systems and global positioning systems are valuable in developing a spatial framework for action, especially in delineating corridors of bio- and ethno-diversity or potential transboundary protected areas.

Experience has shown that many non-governmental organizations (NGOs) have been successful in initiating dialogues with local mountain communities. Success has been based on long-term commitment and integrated (non-sectoral) approaches. International NGOs have been effective partners with national and local NGOs, offering specialized training, facilitation, and mobilization of outside resources. Training can be a powerful tool in mobilizing local environmental knowledge. Such needs are most strongly felt in the development of skills in public communications, leadership and strategic planning.

The possible causes and predicted effects of global climate change have precipitated intense interest in the conservation of species and their habitats. Strategies to address conservation at the global scale require innovative approaches to the acquisition and use of knowledge. While the challenge includes the production of new knowledge, there is also unrealized opportunity in the extensive inventories of raw data provided by contemporary remote sensing technology.

It needs to be emphasized, however, that decisions related to mountain protection need to be taken not only in the seat of political power, but also in accord with the local populations and authorities. An example of reluctance to accept enhanced protection of mountain environments in the early 1990s came from the mountain cantons of Switzerland, which for a long time refused to sign the Alpine Convention, aimed at protecting the remaining natural areas of the Alpine chain from further major development. This is essentially because the populations of these regions did not wish to remain simply as a 'living museum', but desired to have solid economic perspectives for the future. Only after significant changes were made to the original articles did these Swiss cantons ratify the Alpine Convention, thereby clearly signalling that a delicate balance needs to be sought between environmental protection and economic development.

8.2 Adaptation strategies

Adaptation of natural ecosytems to climatic change cannot be achieved without some kind of human intervention, in the form of management of different ecosystems. For certain species which have a reasonably large tolerance to heat and moisture stresses, there may be a degree of natural adaptation, potentially aided by the CO_2 fertilization effect. Reforestation would in some cases be a viable adaptation option, and so would afforestation of abandoned agricultural land.

Freshwater biological systems can be assisted in a number of ways which could help mitigate the impacts of climate change, particularly through the increase and protection of riparian vegetation, and restoring river and stream channels to their natural morphologies. Restoration processes can generally be expected to take several decades. In terms of water resource management, technological measures can be envisaged, including land-use criteria and erosion control, reservoirs and pipelines to increase the availability of freshwater supply, and improvements in the efficiency of water use. Socio-economic options should include direct measures to control water use and land use, as well as indirect measures such as incentives and/or taxes; institutional changes for improved resource management would also need to be envisaged. Specific examples of possible options include rain-fed agriculture with irrigation, water-conserving irrigation practices, enhanced surface water and groundwater management, watershed management, structural and non-structural flood-control management, and reallocation of water resources among water-use sectors and among nations.

For all ecological systems, the reduction of pollution and land-use stresses in the more heavily populated mountain regions, particularly in the valleys, would contribute to removing major stress factors. This could allow plants to adjust more easily to the negative effects of climatic change. An increasingly concentrated effort towards protection and revitalization of freshwater ecosystems in the developed countries has already shown positive achievements, in particular through improvements in the water quality of watersheds. The protection of threatened aquatic habitats such as wetlands has allowed the reintroduction of several endangered verterbrate species; this has been the case in Central Europe, for example, following the transition in the political systems in the early 1990s which has led to increased awareness to environmental problems and to a sharp reduction in toxic pollution levels.

Adaptation options for forests and plants could include the creation of refugia (areas protected by law from human influences), migration corridors and/or assisted migration, and improvements in integrated fire, pest and disease management techniques. This poses problems in many mountain regions of the world, where ecosystems have been so fragmented, and the population density is so high, that some of these options may be impossible to implement. As human populations continue to grow, use of this biologically diverse resource has accelerated, highlighting a strategic role for protected areas that includes both conservation and scientific interests. From a global conservation perspective, protecting representative samples of mountain resources is a high-priority goal. At the same time, scientists are more frequently using protected areas as controls for land-use research and as sites for long-term environmental monitoring. The prospect of global climate change has heightened the value of protected areas as scientific resources (Bailey, 1991). Despite their environmental and scientific value, however, such preserves represent small and fragmented landscapes, making them vulnerable to species migrations and extinctions. In addition, Peterson *et al.* (1990), Peine and Martinka (1992) and Peine and Fox (1995) have shown in various assessment studies that the establishment of refugia in many parts of the world would result in increasing conflicts between different objectives, particularly recreational use and the protection of ecosystems and species. However, it should be stressed that in many instances, the desire to create refugia may be more attractive in terms of attracting tourists than as a purely biodiversity-oriented initiative; it may also lead to conflictual situations with traditional lifestyles (Messerli, personal communication, 1999).

During recent decades, the task of assembling information about species and their habitats has been undertaken by a variety of individuals and institutions. At the international level, inventories being developed by organizational coalitions are critical data sources for strategic conservation planning. The most comprehensive relating to protected areas is maintained by the World Conservation Monitoring Centre (WCMC) as a library of computer maps supported by structured databases and paper files (WCMC, 1992). It represents a global overview of biological diversity that includes protected sites, threatened plant and animal species, and other areas of conservation concern. When organized and presented in an appropriate form, knowledge improves the effectiveness of communications among scientific groups, conservation organizations and managers of natural resources. The potential for a functional network of protected areas is therefore enhanced by the assembly, presentation and application of credible information about the hundreds of sites throughout the world. The concept is consistent with recent recommendations for new approaches to global environmental issues, through the creation of international networks and centres (Carnegie Commission, 1992).

The design of a global network requires that its markets be fully explored to determine the role, functions and applications of information to mountain environments. It seems reasonable to anticipate a diverse clientele, pointing to the need for relevant information that serves both the strategic and the operational requirements of mountain protected areas. As such, the network becomes a contributor to the Sustainable Biosphere Initiative (Lubchenco et al., 1991) through international cooperation in science, education and conservation.

8.3 Policy response

The issues discussed in the following paragraphs are not specific to mountains and uplands, but in view of the fact that many of the environmental problems facing mountains today are of a global nature, their solutions cannot be based solely on regional and local considerations. As a consequence, many of the policy decisions which will be taken in coming decades far from the regions of interest will certainly have a bearing on the pace and severity of environmental change in the world's mountains.

Environmental change presents the decision-maker with numerous sets of challenges. In a set of issues in which there are considerable uncertainties, the policy-makers need to take into account the potential for irreversible damages or costs, and the long time frames involved, i.e. decades to centuries. They must also be aware of the long time lags between emissions and the response of the Earth system to the increased concentrations in greenhouse gases, and the fact that there will be substantial regional variations in impacts. In order to come to terms with global warming, international cooperation is essential but is by no means trivial, because of the wide range of conflicting interests and the extremely heterogeneous income levels in the nations of the world. Because of the long time scales involved, not only in the natural environment but also in terms of the replacement of infrastructure, timely decision-making is of the essence if the negative impacts of change are to be minimized.

Bearing in mind the fact that perhaps the most long-term effects of global environmental change will be felt both directly and indirectly through climate change, it is of interest to have a brief insight into the governing principles of the Framework Convention on Climate Change (FCCC). This convention, along with its sister agreements on Desertification and Biological Diversity, was signed by governments represented at the United Nations Conference on Environment and Development (UNCED) in Rio de Janeiro in June 1992. The FCCC has been ratified since then by over 170 countries, and a number of conferences of the parties to the FCCC have taken place in order to review ways and means of fulfilling the obligations of the various articles of the convention. While the FCCC and the Desertification and Biodiversity Conventions by no means single out mountains as a particular case warranting explicit protection, any long-term concerted policy action aimed at reducing global warming will inevitably have an incidence on the response of mountain and upland environments.

Two articles of the FCCC are of particular relevance to the debate in terms of the environmental and economic principles. Excerpts from Article 2 state that:

> The ultimate objective of the FCCC... is the stabilization of greenhouse gas concentrations in the atmosphere at a level that would prevent dangerous anthropogenic interference with the climate system.... Such a level should be achieved within a time-frame sufficient to allow ecosystems to adapt naturally to climate change, to ensure that food production is not threatened and to enable economic development to proceed in a sustainable manner.

Article 2 of the FCCC poses a number of problems of interpretation, and the manner in which the objectives may be achieved. For example, the objective does not specify what levels of greenhouse gas concentrations may be considered 'safe'. This acknowledges that there is currently no scientific certainty about what a dangerous level would be. Much research will be needed before today's uncertainties are significantly reduced. The convention's objective thus remains meaningful no matter how the science evolves.

The reference to the fact that ecosystems should be able to adapt naturally to climate change (where, in terms of the FCCC, 'change' refers to its anthropogenic component only) is open to a wide range of interpretations. This is because a relatively modest level of climatic change may have little or no impact on certain ecosystems, while others may face extinction at that particular level of change. In this case, the question that arises is whether *all* ecosystems need to be protected, in which case anthropogenic climate change would need to remain within the bounds of natural climatic variability. If only selected ecosystems were to be protected, then it would be essential to determine the thresholds of vulnerability beyond which damage would become irreversible; this is by no means a trivial task, as in many cases these thresholds are not known, because the functioning of plants has for the most part been studied in their current environment and climate.

The manner in which the problem of food security is approached is also open to a wide range of interpretations; it should be borne in mind that agriculture is probably the most climate-sensitive human activity. The IPCC (1996) Report states clearly that global agricultural levels will be able to keep pace with population increases and with climatic change, but that there will be wide regional discrepancies. In this context, should food security concern each individual country, or should it be a concern only at the global level? If the latter, then the poorer nations would depend even more than today on the producing countries of the industrialized world.

But by establishing a framework of general principles and institutions, and by setting up a process through which governments can meet regularly, the FCCC and the other UN conventions have prepared a basis allowing a search for appropriate technical, scientific, legal and institutional solutions. A key benefit of the conventions is that they allow countries to begin discussing an issue even before they all fully agree that it is, in fact, a problem. The conventions are designed to allow countries to weaken or strengthen the treaties in response to new scientific developments. For example, they can agree to take more specific actions, such as reducing emissions of greenhouse gases by a certain amount, through the adoption of amendments or protocols to the FCCC.

The treaties promote action in spite of uncertainty on the basis of a recent development in international law and diplomacy called the 'precautionary principle'. Under traditional international law, an activity has generally not been restricted or prohibited unless a direct causal link between the activity and a particular form of damage can be shown.

But many environmental problems, such as damage to the ozone layer and pollution of the oceans, cannot be confronted if final proof of cause and effect is required. In response, the international community has gradually come to accept the 'precautionary principle', under which activities that threaten serious or irreversible damage can be restricted or even prohibited before there is absolute scientific certainty about their effects.

In Article 3, the governing principles require that:

> the Parties [to the Convention] ... should take precautionary measures to anticipate, prevent or minimize the causes of climate change and mitigate its adverse effects.... Where there are threats of serious or irreversible damage, lack of full scientific certainty should not be used as a reason for postponing such measures, taking into account that policies and measures to deal with climate change should be cost effective so as to ensure global benefits at the lowest possible cost.... To achieve this, such policies and measures should take into account different socio-economic contexts, be comprehensive, cover all relevant sources, sinks and reservoirs of greenhouse gases and adaptation and comprise all economic sectors.... Efforts to address climate change may be carried out cooperatively by interested Parties.

Technological advances have generally increased adaptation options for managed systems. In terms of water resources, for example, these options include more efficient management of existing supplies and infrastructure, and institutional arrangements to limit future demands and to promote conservation. There is also a move towards improved monitoring and forecasting systems for floods and droughts, and construction of new reservoir capacity. In many mountain countries, effective land-use planning can help direct population shifts away from vulnerable locations such as flood plains and steep hillsides. However, many regions of the world currently have limited access to these technologies and appropriate information. The cost-effective use of adaptation strategies will depend among other factors on the availability of financial resources and the degree of technology transfer adapted to the end-user countries. Incorporating environmental change concerns into resource use and development decisions would be a manner by which adaptation strategies could become more efficient.

Many of today's anthropogenically induced environmental problems clearly originate in developed countries, either directly (e.g. greenhouse gas emissions which are leading to global warming) or indirectly (e.g. by delocalizing polluting industries from the rich countries towards the developing world). The basic principle of the FCCC is that the industrialized countries should take the lead in implementing appropriate measures to reduce the amplitude of climatic change and its adverse impacts. Economic growth, social development and environmental protection are interdependent and mutually reinforcing components of sustainable development, which is the framework for international efforts to achieve a higher quality of life worldwide. Responses to environmental change should be coordinated with social and economic development in an integrated manner. Any policy decision should aim at averting the adverse impacts of change, taking into full account the legitimate priority needs of developing countries for the achievement of sustainable development and the eradication of poverty. The FCCC, for example, notes the common but differentiated responsibilities and respective capabilities of all parties to protect the climate system. Specific commitments in the treaty relating to financial and technological transfers apply only to the 24 industrialized countries belonging to the Organisation for Economic Co-operation and Development (OECD). They agree to support climate change activities in developing countries by providing financial support above and beyond any financial assistance they already provide to these countries.

Because a prime contributor to environmental change at large and climatic change in particular is related to energy use, it is in this domain that significant thought must be given to the ways and means of achieving significant energy savings. Such considerations are complicated by the fact that economic and social development is currently governed by the availability and affordability of energy services. Reductions of greenhouse gases and environmental pollutants through cutbacks in energy production and use would lead to intractable economic and development problems which might overshadow the improvements in environmental quality. Wilbanks (1992) states that 'avoiding such a dilemma is the central challenge to energy policy in an era of concern about global environmental change'. However, a set of 'no-regrets' strategies exists, and would allow a balance to be found between environmental concerns and energy-based economic development. The challenge is not to find the best policy today for the entire period between now and the end of the twenty-first century, but to select a coherent strategy and to adjust it over time as new information becomes available. The IPCC (1992, 1994) projects that without policy intervention, there could be significant growth in emissions from the industrial, transportation and commercial and residential sectors. Energy efficiency could increase between 10 and 30 per cent above present levels, through technical conservation measures and improved management practices. Use of existing technologies can help achieve this aim at little or no cost. Because energy use is growing worldwide, even replacing current technology with more efficient technology could still lead to an absolute increase in greenhouse gas emissions in the future.

An extensive array of technologies and policy measures capable of mitigating greenhouse gas emissions is available. There is significant controversy regarding ease and cost of implementation of these technologies and measures, however, due to potential social, institutional, financial, market and legislative barriers to their application and implementation. Without effective policy intervention, the actual contribution of these options to mitigation of emissions may be limited.

Sustainably managed agricultural lands, rangelands and forests can play a key role in reducing current emissions of carbon dioxide, methane and nitrous oxide. Emissions can be reduced, or carbon sequestered, through improved management of agricultural soils, restoration of degraded agricultural lands, establishment of plantations, agroforestry, forest regeneration, and slowing of deforestation. There are significant uncertainties associated with estimating the amount of carbon that can be conserved and/or sequestered.

Energy production from fossil fuels is responsible for the largest portion of greenhouse gas emissions. In 1990, commercial primary energy use was approximately 350 EJ, resulting in the emission of about 6 Gt of carbon. According to IPCC Scenario IS92C, emissions grow to approximately 540–840 EJ by 2025 and 540–2500 EJ by 2100 (IPCC, 1996). The mix of fuels and the geographic sources of emissions are expected to change substantially over this time period. While the likely rate of demand growth is less certain, advanced technologies can significantly reduce both energy demand and resulting emissions.

Over the course of the twenty-first century, the world's energy infrastructure will be replaced completely at least twice, offering opportunities to change the energy system in step with the corresponding investment processes (IPCC, 1996). Advanced technologies can significantly reduce energy demand in all sectors. The most promising options in the industrial sector are more efficient motors, control systems, co-generation, energy cascading and materials recycling. Changes in vehicle design, propulsion systems and fuels, as well as measures to influence driving behaviour, could reduce emissions by 40 per cent in 2025 relative to a high-emission scenario. Projected growth in energy demand in commercial and residential structures could be cut by one-half by 2025 with more efficient heating, cooling, lighting and appliances.

The most promising technology-based approaches for reducing greenhouse gas emissions from energy supply include gas turbine technologies, coal and biomass gasification technologies, production of transportation fuels from biomass, approaches to handling intermittent generation of electricity, wind energy utilization, electricity generation with photovoltaics and solar thermal electric technologies, fuel cells for power generation, and hydrogen as a major new energy carrier, produced first from natural gas and later from biomass and coal, and by electrolysis. Renewable energy sources could provide a substantial portion of foreseeable energy needs of the world over the next century; potentially most promising are biofuels for electricity generation and transportation fuel. Considerable uncertainty exists regarding the costs and rate of market penetration without major policy interventions.

The costs of stabilizing atmospheric concentrations of greenhouse gases at levels and within a time-frame which will prevent 'dangerous anthropogenic interference' with the climate system depend critically on consumption patterns, resource and technology availability, and the choice of policy instruments. Stabilization of *emissions* of greenhouse gases does not imply a stabilization of *concentrations* of these gases in the twenty-first century. Stabilization of concentrations could be achieved only through a drastic reduction of emissions to well below current levels; in its interim report, the IPCC (1994) computed the pathways to stabilization for a number of arbitrary concentration levels, as illustrated in Figure 8.1. Even at very high concentration levels of 750 ppmv of CO_2 at some time in the future, i.e. a threefold increase in atmospheric concentrations compared to pre-industrial levels, the IPCC concludes that emissions

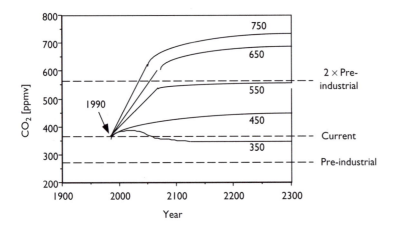

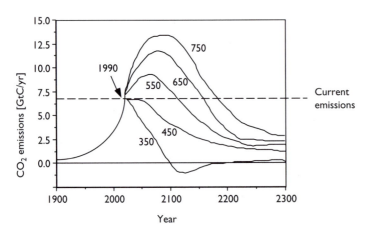

FIGURE 8.1 Arbitrary concentration levels (upper) and possible pathways to stabilization at the selected concentration levels (lower)

would need to be reduced below current levels within a relatively short time-frame.

The choice of abatement paths thus involves balancing the economic risks of rapid abatement now against the risks of delay. Mitigation measures undertaken in a way that capitalizes on other environmental benefits could be cost-effective and enhance sustainable development. Whatever the solutions ultimately chosen, the availability of low-carbon technologies is a prerequisite for, but not a guarantee of, the ability to reduce greenhouse gas emissions at reasonable cost.

Species extinction is a natural part of the evolutionary process. Owing to human activities, however, species and ecosystems are more threatened today than ever before in recorded history. The losses are taking place in tropical forests, where over 50 per cent of the 1.7 million identified species live, as well as in rivers and lakes, deserts and temperate forests, and on mountains and islands. The most recent estimates predict that, at current rates of deforestation, as many as 8 per cent of the Earth's species may disappear by the year 2025. While these extinctions are a consequence of anthropogenically induced environmental change, they also have profound implications for economic and social development. At least 40 per cent of the world's economy and 80 per cent of the needs of the poor are derived from biological resources. In addition, the richer the diversity of life, the greater the opportunity for medical discoveries, economic development and adaptive responses to such new challenges as climate change.

The international community's growing concern over the unprecedented loss of biological diversity has prompted negotiations for a legally binding instrument aimed at reversing this alarming trend. The negotiations were also strongly influenced by the growing recognition throughout the world of the need for a fair and equitable sharing of the benefits arising from the use of genetic resources. These are the key objectives of the UN Convention on Biodiversity, signed as in the case of the FCCC in the context of the UNCED meeting in Rio de Janeiro in 1992. The convention's objectives include 'the conservation of biological diversity, the sustainable use of its components and the fair and equitable sharing of the benefits arising out of the utilization of genetic resources'. The convention is thus the first global, comprehensive agreement to address all aspects of biological diversity, i.e. genetic resources, species and ecosystems. It recognizes that the conservation of biological diversity is 'a common concern of humankind' and an integral part of economic and social development in all countries.

The conservation of biological diversity has ceased to be viewed merely in terms of protecting threatened species or ecosystems. It has emerged as a fundamental part of the move towards sustainable development. Thus the convention introduces a novel approach aimed at reconciling the need for conservation with the concern for development. As in the case of the FCCC, the Biodiversity Convention is based on considerations of equity and shared responsibility. For the first time in the context of biodiversity conservation, an international legal instrument spells out the rights and obligations of its parties concerning scientific, technical and technological cooperation.

In terms of information to the policy-makers, scientists have a key role to play above and beyond simply providing up-to-date results of their research. Scientists need not only to convince politicians of the urgency to act in terms of environmental protection and sustainable economic development, but also to convince society as a whole through accessible results and clearly defined recommendations. As in many aspects of policy-making, it is through the raising of awareness of the general public (i.e. the voters in a democratic system) to various issues that politicians may be moved to consider the issues at hand and attempt to define appropriate response strategies. This may be a more challenging task in certain developing countries, where access to information is often still very difficult, and where the management of mountain resources and environments is less of a priority than health, education, infrastructure and attempting to raise the general standards of living.

8.4 **End note**

In facing up to environmental change, human beings are going to have to think in terms of decades and centuries. Many of the impacts of these profound changes may not become unambiguously apparent for two or three generations. Perhaps the key to success is through long-term economic thinking, based on concepts of sustainable development. Although sustainability is a much-flaunted term today, the common-sense basis for sustainability (i.e. environmental conservation and careful resource use to improve living standards worldwide today, and to provide these resources for future generations) should be seen as the only long-term alternative to current economic trends. The search for sustainability in any form of development presumes that the thresholds of the environmental carrying capacity for a given region are known or can be established on the basis of existing information. For the present time, sustainable economic development can be observed after the fact; in addition, sustainability is a notion which is not necessarily valid for an infinite time, but may change over time as population, technology or the environment shift in response to sustainable policies. The establishment of the goals of sustainable development requires essentially social decisions related to the desirability of establishing a dual environmental–economic system which can survive as long as possible. The real problem here is not to define the goals of sustainability *per se*, but rather to determine the policy implications of what will lead to the establishment of a sustainable system. These considerations, and the large uncertainties associated with them, can only be alleviated to some extent by a consistent application of the precautionary principle mentioned earlier in this chapter.

Many of the policies and decisions related to pollution abatement, climatic change, deforestation or desertification would provide opportunities and challenges for the private and public sectors. A carefully selected set of national and international responses aimed at mitigation, adaptation and improvement of knowledge can reduce the risks posed by environmental change to ecosystems, food security, water resources, human health and other natural and socio-economic systems.

There are large differences in the cost of attempting to address crucial global environmental problems among countries due to their state of economic development, infrastructure choices and natural resource base. International cooperation in a framework of bilateral, regional or international agreements could significantly reduce the global costs of severe environmental stress. If carried out with care, these responses would help to meet the challenge of climate change and enhance the prospects for sustainable economic development for all peoples and nations.

When progress has been made towards attaining some of these global objectives, and the positive effects of implemented policies begin to be perceived, mountain and upland environments will also benefit from these measures. Mountains are unique features of the Earth system in terms of their scenery, their climates, their ecosystems; they provide key resources for human activities well beyond their natural boundaries; and they harbour extremely diverse cultures in both the developing and the industrialized world. The protection of mountain environments against the adverse effects of economic development should be a priority for both today's citizens and the generations to come.

References

Abegg, B. and Froesch, R. 1994: Climate change and winter tourism: impact on transport companies in the Swiss canton of Graübhnden. In Beniston, M. (ed.), *Mountain environments in changing climates*. Routledge, London, 328–340.

Acock, B. and Allen, L.H. 1985: Crop responses to elevated carbon dioxide concentrations. In Strain, B.R. and Cure, J.D. (eds), *Direct effects of increasing carbon dioxide on vegetation*. DOE/ER-0238. US Department of Energy, Washington, DC, 33–97.

Aguado, E., Cayan, D, Riddle, L. and Roos, M. 1992: Changes in the timing of runoff from West Coast streams and their relationships to climatic influences. *Swiss Climate Abstracts*, Special Issue: *International Conference on Mountain Environments in Changing Climates*.

Alcamo, J. (ed.) 1994: *IMAGE 2.0: integrated model of global change*. Kluwer, Dordrecht.

Allen, L.H., Jr, Boote, K.J., Jones, J.W., Jones, P.H., Valle, R.R., Acock, B., Rogers, H.H. and Dahlman, R.C. 1989: Response of vegetation to rising carbon dioxide: photosynthesis, biomass and seed yield of soybean. *Global Biogeochemical Cycles* 1, 1–14.

Allison, I. and Kruss, P. 1977: Estimation of recent climatic change in Irian Jaya by numerical modelling of its tropical glaciers. *Arctic and Alpine Research* 9, 49–60.

Angell, J.K. 1988: Variations and trends in tropospheric and stratospheric global temperatures, 1958–87. *Journal of Climatology* 1, 1296–1313.

Archer S., Schimel D.S. and Holland E.A. 1995: Mechanisms of shrubland expansion: land use, climate or CO_2? *Climatic Change* 29, 91–99.

Armand, A.D. 1992: Sharp and gradual mountain timberlines as a result of species interactions. In Hansen, A.J. and di Castri, F. (eds), *Landscape boundaries: consequences for biotic diversity and ecological flows*. Springer-Verlag, New York, 360–378.

Arnell, N. 1999: The effect of climate change on hydrological regimes in Europe. *Global Environmental Change* 9, 5–23.

Auer, I., Böhm, R. and Mohnl, H. 1990: Die troposphärische Erwarmungsphase die 20. Jahrhunderts im Spiegel der 100-jährigen Messreihe des alpinen Gipfelobservatoriums auf dem Sonblick. CIMA '88. Congresso Ventesimo Meteorologica Alpina (Sestola, Italy). Italian Meteorological Service, Rome.

Aulitzky, H., Turner, H. and Mayer, H. 1982: Bioklimatische Grundlagen einer standortgemässen Bewirtschaftung des subalpinen Lärchen – Arvenwaldes, *Eidgenössische Anstalt für forstliches Versuchswesen Mitteilungen* 58, 325–580.

Baig, M.N. and Tranquillini, W. 1980: The effects of wind and temperature on cuticular transpiration of *Picea abies* and *Pinus cembra* and their significance in desiccation damage at the alpine treeline. *Oecologia* 47, 252–256.

Bailey, R.G. 1991: Design of ecological networks for monitoring global change. *Environmental Conservation* 18, 173–175.

Balteanu, D., Ozenda, P., Huhn, M., Kershner, H., Tranquillini, W. and Bartenschluyer, S. 1987: Impact analysis of climatic change in the central European mountain ranges, *European Workshop on Interrelated Bioclimatic and Land Use Changes*, Volume G. Noordwijkerhout, The Netherlands.

Banzon, V., de Franceschi, G. and Gregori, G. 1992: The mathematical handling and analysis of non-homgeneous and incomplete multivariate historical data series. In Frenzel, B., Pfister, C. and Glaeser, B. (eds), *European climate reconstructed from documentary data: methods and results*. Gustav Fischer, Stuttgart, 137–151.

Barker, M.L. 1982: Traditional landscape and mass tourism in the Alps. *Geographical Review* 72, 395–415.

Barnola, J.-M., Pimienta, P., Raynaud, D. and Korotkevich, Y.S. 1991: CO_2-climate relationship as deduced from the Vostok ice core: a re-examination based on new measurements and on a re-evaluation of the air dating. *Tellus* 43, 83–90.

Barry, R.G. 1986: Mountain climate data for long-term ecological research. In *Proceedings of International Symposium on the Qinghai-Xizang Plateau and Mountain Meteorology*. Science Press, Beijing, 170–187.

Barry, R.G. 1992a: *Mountain weather and climate*. Routledge, London.

Barry, R.G. 1992b: Mountain climatology and past and potential future climatic changes in mountain regions: a review, *Mountain Research and Development* 12, 71–86.

Barry, R.G. 1994: Past and potential future changes in mountain environments: a review. In Beniston, M. (ed.), *Mountain environments in changing climates*. Routledge, London, 3–33.

Barry, R.G. and Van Wie, C.C. 1974: Topo- and microclimatology in alpine areas. In *Arctic and Alpine Environment*. Methuen, London, 73–83.

Bates, G.T., Hostettler, S.W. and Giorgi, F. 1995: Two-year simulation of the Great Lakes region with a coupled modeling system. *Monthly Weather Review* **123**, 1505–1522.

Baumgartner, T.R., Michaelsen, J., Thompson, L.G., Shen, G.T., Soutar, A. and Casey, R.E. 1989: The recording of inter-annual climatic change by high resolution natural systems: tree-rings, coral bands, glacial ice layers and marine varves. In Peterson, D. (ed.), *Climatic change in the eastern Pacific and western Americas*, American Geophysical Union, Washington, DC, 1–14.

Beaver, P.D. and Purrington, B.L. (eds) 1984: *Cultural adaptation to mountain environments.* University of Georgia Press, Athens (USA).

Bellwald, W. 1992: Drei spätneolithisch/frühbronze-zeitliche Pfeilbogen aus dem Gletschereis am Lötschenpass. *Archäologie der Schweiz* **15**, 166–171.

Beniston, M. (ed.) 1994: *Mountain environments in changing climates.* Routledge, London.

Beniston, M. 1997: *From turbulence to climate.* Springer, Heidelberg.

Beniston, M., Fox, D.G., Adhikary, S. *et al.* 1996: *The impacts of climate change on mountain regions.* Second Assessment Report of the Intergovernmental Panel on Climate Change (IPCC), Chapter 5, Cambridge University Press, Cambridge, 191–213.

Beniston, M., Haeberli, W., Hoelzle, M. and Taylor, A. 1998: On the potential use of glacier and permafrost observations for verification of climate models. *Annals of Glaciology* **25**, 400–406.

Beniston, M., Ohmura, A., Rotach, M., Tschuck, P., Wild, M. and Marinucci, M.R. 1995: Simulation of climate trends over the Alpine Region: development of a physically-based modeling system for application to regional studies of current and future climate. Final Scientific Report No. 4031-33250 to the Swiss National Science Foundation, Bern, Switzerland.

Beniston, M. and Price, M. 1992: Climate scenarios for the Alpine regions: A collaborative effort between the Swiss National Climate Program and the International Center for Alpine Environments. *Environmental Conservation* **19**, 360–363.

Beniston, M. and Rebetez, M. 1996: Regional behavior of minimum temperatures in Switzerland for the period 1979–1993. *Theoretical and Applied Climatology* **53**, 231–243.

Beniston, M., Rebetez, M., Giorgi, F. and Marinucci, M. R. 1994: An analysis of regional climate change in Switzerland. *Theoretical and Applied Climatology* **49**, 135–159.

Beniston, M., Rotach, M., Tchuck, P., Wild, M. and Ohmura, A. 1996b: Feedbacks between mountains and climate: numerical studies with global and regional models in the context of the CLEAR Project of the Priority Program on the Environment. Final Scientific Report No. 5001-035179 to the Swiss National Science Foundation, Bern, Switzerland.

Bennett, S., Brereton, R., Mansergh, I., Berwick, S., Sandiford, K. and Wellington, C. 1991: *The potential effect of the enhanced climate change on selected Victorian fauna.* Technical Report Series No. 123, Arthur Rylah Institute, Heidelberg, Victoria, Australia.

Betancourt, J.L., Pierson, E.A., Aasen Rylander, K., Fairchild-Parks, J.A. and Dean, J.S. 1993: Influence of history and climate on New Mexico piñon–juniper woodlands. In Aldon, E.F. and Shaw, D.W. (eds), *Managing piñon–juniper ecosystems for sustainability and social needs.*

Gen. Tech. Rep. RM-236, Fort Collins, CO, USDA Forest Service, Rocky Mountain Forest and Range Experimental Station, 42–62.

Bie, S.W. and Imevbore, A.M.A. 1995: Executive summary. In Bie, S.W. and Imevbore, A.M.A. (eds), *Biological diversity in the drylands of the world.* United Nations, New York, 5–21.

Bjerrum, L. and Jfrstad, F. 1968: *Stability of rock slopes in Norway,* Norwegian Geotechnical Institute Publication 79, 1–11.

Bolin, B. 1950: On the influence of the earth's orography on the general character of the westerlies. *Tellus* **2**, 184–195.

Bortenschlager, S. 1993: Das höchst gelegene Moor der Ostalpen 'Moor am Rofenberg' 2760 m. PhD dissertation, Department of Botany, University of Innsbruck, Austria.

Botkin, D.B., Janak J.F. and Wallis J.R. 1972: Some ecological consequences of a computer model of forest growth. *Journal of Ecology* **60**, 849–871.

Bowler, J.M., Hope, G.S., Jennings, J.N., Singh, G. and Walker, D. 1976: Late Quaternary climates of Australia and New Zealand. *Quaternary Research* **6**, 359–399.

Boza, M.A 1992: *Parques nacionales Costa Rica.* Editorial Incafo, Heredia, Costa Rica.

Braat, L.C. and van Lierop, W. F. (eds) 1987: *Economic-ecological modeling.* Studies in Regional Science and Urban Economics, 16, Kluwer, Dordrecht.

Bradshaw, R.H.W. and Zackrisson, O. 1990: A two thousand year history of a northern Swedish boreal forest stand. *Journal of Vegetation Science* **1**, 519–528.

Bravo, R.E., Canadas-Cruz, L., Estrada, W. *et al.* 1988: The effects of climatic variations on agriculture in the Central Sierra of Ecuador. In Parry, M.L., Carter, T.R. and Konijn, N.T. (eds), *The impact of climate variations on agriculture*: vol. 2, *Assessments in semi-arid regions.* Kluwer, Dordrecht, 381–493.

Breiling, M. and Charamza, P. 1994: Localizing the threats due to climate change in mountain environments. In Beniston, M. (ed.), *Mountain environments in changing climates,* Routledge, London, 341–365.

Broccoli, A.J. and Manabe, S. 1992: The effect of orography on midlatitude northern hemisphere dry climates. *Journal of Climate* **5**, 1181–1201.

Broecker, W.S., Peteet, D. and Rind, D. 1985: Does the ocean-atmosphere system have more than one stable mode of operation? *Nature* **315**, 21–26.

Brubaker, L.B. 1986: Responses of tree populations to climate change. *Vegetatio* **67**, 119–130.

Bruijnzeel, L.A. and Critchley, W.R.S. 1994: *Environmental impacts of logging moist tropical forests.* IHP Humid Tropics Programme, Series 7, UNESCO.

Bryner, G. 1991: Implementing global environmental agreements. *Policy Studies Journal* **19**, 103–114.

Bugmann, H. 1994: On the ecology of mountain forests in a changing climate: a simulation study. PhD thesis No. 10638, Swiss Federal Institute of Technology (ETH), Zürich, Switzerland, 268.

Bugmann, H. and Fischlin, A. 1994: Comparing the behaviour of mountainous forest succession models in a changing climate. In Beniston, M. (ed.), *Mountain environments in changing climates,* Routledge, London, 204–219.

Bultot, F., Gellens, D., Schädler, B. and Spreafico, M. 1992: Impact of climatic change, induced by the doubling of the atmospheric CO_2 concentration, on snow cover characteristics: case of the Broye drainage basin in Switzerland. *Swiss Climate Abstracts*, Special Issue: *International Conference on Mountain Environments in Changing Climates*.

Burri, K. 1995: *Schweiz*. Lehrmittelverlag des Kantons Zürich.

Burrows, C.J. 1990: *Processes of vegetation change*. Unwin Hyman, London.

Callaghan, T.V. and Jonasson, S. 1994: Implications from environmental manipulation experiments for Arctic plant biodiversity changes. In Chapin, F.S. and Körner, C. (eds), *Arctic and alpine biodiversity: patterns, causes and ecosystem consequences*, Springer-Verlag, Stuttgart.

Campos M., Sanchez, A. and Espinoza, D. 1996: *Adaptation of hydropower generation in Costa Rica and Panama to climate change*. Central American Project on Climate Change, Springer-Verlag, Stuttgart.

Carnegie Commission 1992: *International environmental research and assessment: proposals for better organization and decision making*. Carnegie Commission on Science, Technology and Government, New York.

Carter, T.R. and Parry, M.F. 1994: Evaluating the effects of climatic change on marginal agriculture in upland areas. In Beniston, M. (ed.), *Mountain environments in changing climates*. Routledge, London, 405–421.

Carver, M. and Nakarmi, G. 1995: The effect of surface conditions on soil erosion and stream suspended sediments. In Schreir, H., Shah, P.B. and Brown, S. (eds), *Challenges in resource dynamics in Nepal: processes, trends and dynamics in middle mountain watersheds Proceedings of an ICIMOD Workshop* (International Center for Integrated Mountain Development). Kathmandu, Nepal, 155–162.

Cess, R.D. and 31 co-authors 1990: Intercomparison and interpretation of climate feedback mechanisms in 19 atmospheric general circulation models. *Journal of Geophysical Research* **95** (D10), 16601–16615.

Chalise, S. 1994: Mountain environments and climate change in the Hindu Kush–Himalayas. In Beniston, M. (ed.), *Mountain environments in changing climates*. Routledge, London, 382–404.

Chalise, S. 1997: *Mountain tourism in Nepal*. Mountain Forum Internet Publication on http://www.mtnforum.org/

Charlesworth, J.K. 1957: *The quaternary era*. Arnold, London.

Chen, J.Y. and Ohmura, A. 1990: On the influence of alpine glaciers on runoff. In Lang, H. and Musy, A. (eds), *Hydrology in mountainous regions. I. Hydrological measurements. The water cycle*. IAHS Press, Wallingford, UK, 127–135.

Chen, S.-C., Roads, J., Juang, H.H.-M. and Kanamitsu, M. 1994: California precipitation simulation in the NMC nested spectral model: the January 1993 event. *Proceedings of Symposium to Share Weather Pattern Knowledge*, June 1994, Rocklin, California, USA.

Chippendale, G.M. 1979: *The southwestern corn borer, Diatraea grandiosella: case history of an invading insect*. Research Bulletin 1031, University of Missouri Agricultural Experiment Station, Columbia.

Clark J.S. 1988: Effect of climate change on fire regimes in northwestern Minnesota. *Nature* **334**, 233–235.

Clark, J.S. 1990: Fire and climate change during the last 750 years in northwest Minnesota. *Ecological Monthly* **60**, 135–159.

Cliff, A.D. and Ord, J.K. 1981: *Spatial autocorrelation*. Pion, London.

CLIMAP Project Members 1976: The surface of the Ice Age earth. *Science* **191**, 1131–1137.

CLIMAP Project Members 1981: *Seasonal reconstruction of the earth's surface at the last glacial maximum*. Geological Society of America, Map Chart Series, MC-36.

CLIVAR 1997: *Climate variability and predictability, implementation plan*. World Meteorological Organization, Geneva.

COHMAP Members 1988: Climatic changes of the last 18,000 years: observations and model simulation. *Science* **241**, 1043–1052.

Collins, D.N. 1989: Hydrometeorological conditions, mass balance and runoff from alpine glaciers. In Oerlemans, J. (ed.), *Glacier fluctuations and climate change*. Kluwer, Dordrecht, 305–323.

Copeland, J.H., Chase, T., Baron, J., Kittel, T.G.F. and Pielke, R.A. 1994: Impacts of vegetation change on regional climate and downscaling of GCM output to the regional scale. *Proceedings of the 32nd Hanford Symposium on Health and the Environment*, October 1993, Richland, WA.

Copeland, J.H., Pielke, R.A. and Kittel, T.G.F. 1996: Potential climatic impacts on vegetation change: a regional modeling study. *Journal of Geophysical Research* **101**, 7409–7418.

Cress, A., Davies, H.C., Frei, C., Lüthi, D. and Schaer, C. 1994: *Regional climate simulations in the Alpine region*. LAPETH-32, Atmospheric Sciences ETH, Zürich.

Cress, A., Majewski, D., Podzun, R. and Renner, V. 1995: Simulation of European climate with a limited area model. Part I: Observed boundary conditions. *Contributions to Atmospheric Physics* **72**, 31–52.

Cumming, S.G. and Burton, P.J. 1996: Phenology-mediated effects of climatic change on some simulated British Columbia forests. *Climatic Change* **34**, 213–222.

Cure, J.D. 1985: Carbon dioxide doubling responses: A crop survey. In Strain, B.R. and Cure, J.D. eds, *Direct effects of increasing carbon dioxide on vegetation*. DOE/ER-0238, US Department of Energy, Washington, DC, 33–97.

Dahlback, A. 1989: Biological UV doses and the effect of an ozone layer depletion. *Photochemistry and Photobiology* **49**, 621–625.

Dankelman, I. and Davidson, J. 1991: Land: women at the center of the food crisis. In Sontheimer, S. (ed.), *Women and the environment*. Monthly Review Press, New York.

Danks, H.V. 1991: Winter habitats and ecological adaptations for winter survivals. In Lee, R.E. and Denlinger, D.L. (eds), *Insects at low temperature*. Chapman & Hall, New York.

David, F. 1993: Altitudinal variation in the response of vegetation to late-glacial climatic events in the northern French Alps, *New Phytologist* **125**, 203–220.

Davis, M.B. 1986: Climatic instability, time lags and community disequilibrium. In Diamond, J. and Case, T.J. (eds), *Community ecology*. Harper & Row, New York, 269–284.

DeWalt, K.M. 1993: Nutrition and the commercialisation of agriculture: ten years later. *Social Science and Medicine* **36**, 1407–1416.

Diaz, H.F. and Bradley, R.S. 1997: Temperature variations during the last century at high elevation sites. *Climatic Change* **36**, 253–279.

Diaz, H.F. and Graham, N.E. 1996: Recent changes in tropical freezing heights and the role of sea-surface temperature. *Nature* **383**, 152–155.

Diemer, M., Körner, C. and Prock, S. 1992: Leaf life spans in wild perennial herbaceous plants: a survey and attempts at a functional interpretation. *Oecologia* 89, 10–16.

Divya, G. and Mehrotra, R. 1995: Climate change and hydrology with emphasis on the Indian subcontinent. *Hydrological Sciences Journal* 40, 231–241.

Downing, T.E. 1992: *Climate change and vulnerable places: global food security and country studies in Zimbabwe, Kenya, Senegal and Chile.* Environmental Change Unit, Oxford.

Downton, M.W., Stewart, T.R. and Miller, K.A. 1988: Estimating historical heating and cooling needs: per capita degree days. *Journal of Applied Meteorology* 27, 84–90.

Edwards, P.J. and Grubb, P.J. 1988: Studies of mineral cycling in a New Guinean montane forest; II: The production and disappearance of litter. *Journal of Ecology* 65, 971–992.

Ekhart, E. 1948: De la structure de l'atmosphère dans la montagne. *La Météorologie* 3, 3–26.

Elliot, K.J. and Swank, W.T. 1994: Impacts of drought on tree mortality and growth in a mixed hardword forest. *Journal of Vegetation Science* 5, 229–236.

Epstein, P.L. 1992: Cholera and the environment. *Lancet* 339, 1167–1168.

Eybergen, J. and Imeson, F. 1989: Geomorphological processes and climate change. *Catena* 16, 307–319.

FAO 1986: *Wildland fire management terminology.* Food and Agriculture Organization, Rome.

Feacham, R., Kjellstrom, T., Murray, C.T.L., Over, M. and Phillips, M.A. 1980: *Health aspects of excreta and sullage management.* Appropriate Technology for Water Supply and Sanitation Series, World Bank, Washington, DC.

Fischer, G., Frohberg, K., Keyzer, M.A., Parikh, K.S. and Tims, W. 1990: *Hunger: beyond the reach of the invisible hand*.: International Institute for Applied Systems Analysis, Laxenburg, Austria, Food and Agriculture Project.

Fischlin, A. 1995: Assessing sensitivities of forests to climate change: experiences from modelling case studies. In Guisan, A., Holten, J.I., Spichiger, R. and Tessier, L. (eds), *Potential ecological impacts of climate change in the Alps and Fennoscandian Mountains.* Geneva University Press, Geneva, 145–149.

Fitzharris, B.B., Allison, I., Braithwaite, R.J., Brown, J., Foehn, P., Haeberli, W., Higuchi, K., Kotlyakov, V.M., Prowse, T.D., Rinaldi, C.A., Wadhams, P., Woo, M.K. and Youyu Xie 1996: The cryosphere: changes and their impacts. In *Second Assessment Report of the Intergovernmental Panel on Climate Change (IPCC)*, Chapter 5, Cambridge University Press, Cambridge, 241–265.

Flenley, J.R. 1979: The late quaternary vegetational history of the equatorial mountains. *Progress in Physical Geography* 3, 488–509.

Fliri, F. 1975: *Das Klima der Alpen im Raume von Tirol.* Universitätsverlag Wagner, Innsbruck and Mhnchen.

Fliri, F. 1982: *Tirol-Atlas – D. Klima.* Universitätsverlag Wagner, Innsbruck.

Flohn, H. 1968: *Contributions to a meteorology of the Tibetan Highlands.* Atmospheric Physics Paper 130, Department of Atmospheric Sciences, Colorado State University, Fort Collins.

Föhn, P. 1991: Les hivers de demain seront-ils blancs comme neige ou vert comme les prés?, WSL/FNP (ed.), *Argument de la Recherche* 3, 3–12.

Fosberg, M.A. 1990: Global change – a challenge to modeling. In Dixon, R.K., Meldahl, R.S., Ruark, G.A. and Warren, W.G. (eds), *Process modeling of forest growth responses to environmental stress.* Timber Press, Portland, OR, 3–8.

Franklin, J.F., Shugart, H.H. and Harmon, M.E. 1987: Tree death as an ecologcal process. *BioScience* 38, 550–556.

French, H.M. 1996: *The periglacial environment.* Longman, London.

Fritts, H.C. 1976: *Tree rings and climate.* Academic Press, New York.

Fuentes, E.R. and Castro, M. 1982: Problems of resource management and land use in two mountain regions of Chile. In di Castri, F., Baker, G. and Hadley, M. (eds), *Ecology in practice*, Tycooly Press, Dublin, 315–330.

Galloway, R.W. 1988: The potential impact of climate changes on Australian ski fields. In Pearman, G.I., (ed.), *Greenhouse planning for climate change*, CSIRO, Aspendale, Australia, 428–437.

Gamper, M. and Suter, J. 1982: Postglaziale Klimageschichte der Alpen. *Geographica Helvetica* 37, 105–114.

Garfinkel, H.L. and Brubaker, L.B. 1980: Modern climate–tree-ring relations and climatic reconstruction in sub-arctic Alaska. *Nature* 286, 872–873.

Garr, C.E. and Fitzharris, B.B. 1994: Sensitivity of mountain runoff and hydro-electricity to changing climate. In Beniston, M. (ed.), *Mountain ennvironment in changing climates*. Routledge, London, 336–381.

Gates, M.D. 1993: *Climate change and its biological consequences*. Snauer Associates, Sunderland, MA.

Gates, W.L. 1992: AMIP: The atmospheric model intercomparison project. *Bulletin of the American Meteorological Society* 73, 1962–1970.

Geiser, F. and Broome, L. 1993: The effects of temperature on the pattern of torpor in a marsupial hibernator. *Journal of Comparative Physiology* 163, 133–137.

Giorgi, F., Brodeur, C.S. and Bates, G.T. 1994: Regional climate change scenarios over the United States produced with a nested regional climate model. *Journal of Climatology* 7, 375–399.

Giorgi, F., Hurrell, J., Marinucci, M. and Beniston, M. 1997: Height dependency of the North Atlantic Oscillation Index: observational and model studies. *Journal of Climatology* 10, 288–296.

Giorgi, F., Marinucci, M.R. and Bates, G.T. 1993: Development of a second-generation regional climate model. Part I: Boundary-layer and radiative transfer processes. *Monthly Weather Review* 121, 2794–2813.

Giorgi, F. and Mearns, L.O. 1991: Approaches to the simulation of regional climate change: a review. *Reviews of Geophysics* 29, 191–216.

Glantz, M.H. (ed.) 1988: *Societal responses to regional climatic change.* Westview Press, Boulder, CO.

Glantz, M.H., Price, M.F. and Krenz, M.E. (eds) 1990: *On assessing winners and losers in the context of global warming.* National Center for Atmospheric Research, Boulder, CO.

Gleick, P.H. 1986: Methods for evaluating the regional hydrologic impacts of global climatic changes, *Journal of Hydrology* 88, 97–116.

Gleick, P.H. 1987a: Regional hydrologic consequences of increases in atmospheric CO_2 and other trace gases. *Climate Change* 10, 137–161.

Gleick, P.H. 1987b: The development and testing of a water balance model for climate impact assessment: modelling the Sacramento Basin. *Water Resources Research* 23, 1049–1061.

Gleick, P. 1994: Effects of climate change on shared water resources. In Gleick, P. (ed.), *Water in crisis*. Oxford University Press, Oxford.

Goliber, T.J. 1985: Sub-Saharan Africa: population pressures and development. *Population Bulletin* **40**, 1–47.

Gonzalez, J.A. 1985: El potencial agua en algunas plantas de altura y el problema del stress hídrico en alta montana. *Lilloa* **36**, 167–172.

Goto, N, Sakoto, A. and Suzuki, M. 1994: Modelling of soil carbon dynamics as part of the carbon cycle in terrestrial ecosystems. *Ecological Modelling* **19**, 202–243.

Goudriaan, J. and Unworth, H.M. 1990: Implications of increasing carbon dioxide and climate change for agricultural productivity and water resources. In Heichel, G. H. (ed.), *Impact of carbon dioxide trace gases and climate changes on global agriculture*. American Society of Agronomy Special Publication 53, Madison, WI, 111.

Government of Tanzania 1979: *Report of Population Census 1978*. Bureau of Statistics, Dar-es-Salaam.

Govi, M. 1990: Conférence spéciale: *Mouvements de masse récents et anciens dans les Alpes italiennes. Proceedings of the Fifth Symposium on Landslides*, Lausanne **3**, 1509–1514.

Grabherr, G., Gottfried, M. and Pauli, H. 1994: Climate effects on mountain plants. *Nature* **369**, 448.

Graham R.L., Turner, M.G. and Dale, V.H. 1990: How increasing CO_2 and climate change affect forests. *BioScience* **40**, 575–587.

Granier, C., Muller, J.F. and Hao, W.M. 1998: The impact of biomass burning on the global budget of ozone and ozone precursors. *Abstracts of the International Workshop on Biomass Burning and its Inter-relationships with the Climate System*. Wengen, Switzerland, September 1998.

Grassl, H. 1994: The Alps under local, regional and global pressures. In Beniston, M. (ed.), *Mountain environments in changing climates*. Routledge, London, 34–41.

Graumlich, L.J. 1991: Subalpine tree growth, climate and increasing CO_2: an assessment of recent growth trends. *Ecology* **72**, 1–11.

Graumlich, L.J. 1993: A 1000-year record of temperature and precipitation in the Sierra Nevada. *Quaternary Research* **39**, 249–255.

Graumlich, L.J. and Brubaker, L.B. 1986: Reconstruction of annual temperature (1590–1979) for Longmire, Washington, derived from tree rings. *Quaternary Research* **25**, 223–234.

Graybill, D.A. and Idso, S.B. 1993: Detecting the aerial fertlization effect of atmospheric CO_2 enrichment in tree-ring chronologies. *Global Biogeochemical Cycles* **7**, 81–95.

Green, K., Broome, L. and Osborne, W. 1997: *Snow: a natural history, an undertain future*. Australian Alps National Parks Publication, Canberra.

Grimm, E.C. 1983: Chronology and dynamics of vegetation change in the prairie and woodland region of southern Minnesota, USA. *New Phytologist* **93**, 311–350.

Grötzbach, E.F. 1988: High mountains as human habitat, In Allan, N.J.R., Knapp, G.W. and Stadel, C. (eds), *Human impact on mountains*. Rowman & Littlefield, Totowa, NJ, 24–35.

Grove, J.M. 1987: *The Little Ice Age*. Methuen, London.

Grubb, P.J. and Tanner, E.V. 1976: The montane forests and soils of Jamaica: a reassessment. *Journal of the Arnold Arboretum* **57**, 313–368.

Guillet, D. 1983: Toward a cultural ecology of mountains: the Central Andes and the Himalaya compared. *Current Anthropology* **24**, 561–574.

Guisan, A. and Holten, J.I. 1995: Impacts of climate change on mountain ecosystems: future research and monitioring needs In Guisan, A., Holten, J.I, Spichiger, R., Tessier, L. (eds), *Potential ecological impacts of climate change in the Alps and Fennoscandian mountains*. Geneva University Press, Geneva, 179–185.

Guisan, A., Holten, J., Spichiger, R. and Tessier, L. (eds) 1995a: *Potential impacts of climate change on ecosystems in the Alps and Fennoscandian mountains*. Annex Report to the IPCC Working Group II Second Assessment Report. Publication Series of the Geneva Conservatory and Botanical Gardens, University of Geneva, Switzerland.

Guisan, A., Theurillat, J.-P. and Spichiger, R. 1995b: Effects of climate change on Alpine plant diversity and distribution: the modelling and monitoring perspectives. In Guisan, A., Holten, J.I., Spichiger, R. and Tessier, L. (eds), *Potential ecological impacts of climate change in the Alps and Fennoscandian mountains*. Geneva University Press, Geneva, 129–137.

Gunderson, C.A. and Wullschleger, S.D. 1994: Photosynthetic acclimation in trees to rising atmospheric CO_2: a broader perspective. *Photosynthetic Research* **39**, 369–388.

Gunn, J.M. and Keller, W. 1990: Biological recovery of an acid lake after reduction in industrial emissions of sulphur. *Nature* **345**, 431–433.

Gyalistras, D., von Storch, H., Fischlin, A. and Beniston, M. 1994: *Linking GCM-simulated climatic changes to ecosystem models: case studies of statistical downscaling in the Alps. Climate Research* **4**, 167–189.

Haeberli, W. 1985: *Creep of mountain permafrost: internal structure and flow of Alpine rock glaciers*. Internal Report 77 of the Institute for Hydraulic Constructions (VAW), Swiss Federal Institute of Technology, Zürich.

Haeberli, W. 1990: Glacier and permafrost signals of twentieth century warming. *Annals of Glaciology* **14**, 99–101.

Haeberli, W. 1994: Accelerated glacier and permafrost changes in the Alps. In Beniston, M. (ed.), *Mountain environments in changing climates*. Routledge, London, 91–107.

Haeberli, W. 1995: Glacier fluctuations and climate change detection – operational elements of a worldwide monitoring strategy. *WMO Bulletin* **44**, 23–31.

Haeberli, W. and Beniston, M. 1998: Climate change and its impacts on glaciers and permafrost in the Alps. *Ambio* **27**, 258–265.

Haeberli, W. and Funk, M. 1991: Borehole temperatures at the Colle Gnifetti core-drilling site (Monte Rosa, Swiss Alps). *Journal of Glaciology* **37**, 37–46.

Haeberli, W., Muller, P., Alean, J. and Bösch, H. 1990: Glacier changes following the Little Ice Age: a survey of the international data base and its perspectives. In Oerlemans, J. (ed.), *Glacier fluctuations and climate*. Reidel, Dordrecht, 77–101.

Halpin, P.N. 1994: Latitudinal variation in montane ecosystem response to potential climatic change. In Beniston, M. (ed.), *Mountain ecosystems in changing climates*, Routledge, London, 180–203.

Hanawalt, R.B. and Whittaker, R.H. 1976: Altitudinally coordinated patterns of soil and vegetation in the San Jacito Mountains, California. *Soil Science* **121**, 114–124.

Hansen-Bristow, K.J., Ives, J.D. and Wilson, J.P. 1988: Climatic variability and tree response within the forest–alpine tundra ecotone. *Annals of the Association of American Geographers* **78**, 505–519.

Hastenrath, S. and Kruss, P.D. 1992: The dramatic retreat of Mount Kenya's glaciers 1963–87: greenhouse forcing. *Annals of Glaciology* **16**, 127–133.

Hastenrath, S. and Rostom, R. 1990: Variations of the Lewis and Gregory Glaciers, Mount Kenya, 1978–86–90. *Erdkunde* **44**, 313–317.

Haxeltine, A. and Prentice, I.C. 1996: A general model for the light-use efficency of primary production. *Functional Ecology* **29**, 347–372.

Haxeltine, A., Prentice, I.C. and Cresswell, I.D. 1996: A coupled carbon and water flux model to predict vegetation structure. *Journal of Vegetation Science* **7**, 651–666.

Heal, O.W., Menault, J.P. and Steffen, W.L. (eds) 1993: *Towards a global terrestrial observing system (GTOS): detecting and monitoring change in terrestrial ecosystems.* IGBP Global Change Report 26, Fontainebleau.

Hedberg, O. 1964: The phytogeographical position of the afroalpine flora. *Recent Advances in Botany* 914–919.

Hedberg, O. 1995: *Features of Afro-alpine plant ecology.* ABC O Ekblad, Vastervik.

Held, I.M. 1983: *Stationary and quasi-stationary eddies in the extratropical troposphere: theory.* Academic Press, New York, 127–168.

Henderson-Sellers, A. and McGuffie, K. 1987: *A climate modeling primer.* Wiley, New York.

Hengeveld, R. 1990: Theories on species responses to variable climates. In Boer, M.M. and de Groot, R.S. (eds), *Landscape–ecological impact of climatic change.* IOS Press, Amsterdam, 274–293.

Heuberger, H. 1966: Gletschergeschichtliche Untersuchungen in den Zentralalpen zwischen Sellrain und Oetztal. *Wissenschaft Alpenvereinhefte* **20**, 126.

Hirakuchi, H. and Giorgi, F. 1995: Multi-year present-day and $2 \times CO_2$ simulations of monsoon climate over eastern Asia and Japan with a regional climate model nested in a general circulation model. *Journal of Geophysical Research* **100**, 4327–4338.

Holdridge, L.R. 1947: Determination of world plant formations from simple climatic data. *Science* **105**, 367–368.

Holtmeier, F.-K. 1986: Die obere Waldgrenze unter dem Einfluß von Klima und Mensch. *Abhandlungen aus dem Westfälischen Museum für Naturkunde zu Münster* **48**, 395–412.

Holtmeier, F.-K. 1994: Ecological aspects of climatically caused timberline fluctuations: review and outlook. In Beniston, M. (ed.), *Mountain environments in changing climates*, Routledge, London, 220–233.

Hope, G.S., Peterson, J.A., Radok, U. and Allison, I. (eds) 1976: *The equatorial glaciers of New Guinea.* A.A. Balkema, Rotterdam.

Houghton, J. 1984: *The global climate.* Cambridge University Press, Cambridge.

Huggett, R.J. 1995: *Geoecology.* Routledge, London.

Hulme, M. 1997: Global warming. *Progress in Physical Geography* **21**, 446–453.

Humboldt, A. von 1817: *De distributionae geographica plantarum.* Libraria Graeco-Latino-Germanica, Paris.

Huntley, B. 1991: How plants respond to climate change: migration rates, individualism and the consequences for plant communities. *Annals of Botany* **67**, 15–22.

Hurni, H. and Stähli, P. 1982: Das Klima von Semien. In Hurni, H. (ed.), *Hochgebirge von Semien: Äethiopen*, Volume 2, *Kaltzeit bis zur Gegenwart.* Geographica Bernensia 13. University of Bern, 50–82.

Hurrell, J.W. 1995: Decadal trends in the North Atlantic Oscillation regional temperatures and precipitation. *Science* **269**, 676–679.

Hurrell, J.W. and van Loon, H. 1997: Decadal variations in climate associated with the North Atlantic Oscillation. *Climatic Change* **36**, 301–326.

Iafiazova, R.K. 1997: Climate change impact on mud flow formation in Trans-Ili Alatay mountains. *Hydrometeorology and Ecology* **3**, 12–23 (in Russian).

IAHS (ICSI)/UNEP/UNESCO 1989: *World Glacier Inventory – Status 1988*, ed. Haeberli, W., Bösch, H., Scherler, K., Østrem, G. and Wallén, C.C. Nairobi.

IAHS (ICSI)/UNEP/UNESCO 1993: *Fluctuations of Glaciers 1985–1990*, ed. Haeberli, W. and Hoelzle, M. Paris.

IAHS (ICSI)/UNEP/UNESCO 1994: *Glacier Mass Balance Bulletin* no. 3, ed. Haeberli, W., Hoelzle, M. and Bösch, H. World Glacier Monitoring Service, ETH, Zürich.

Innes, J.L. 1994: Design of an intensive monitoring system for Swiss forests. In Beniston, M. (ed.), *Mountain environments in changing climates.* Routledge, London, 281–298.

Innes, J.L. 1998: The impacts of climatic extremes on forests: an introduction. In Beniston, M. and Innes, J.L. (eds), *The impacts of climate variability on forests*, Springer-Verlag, Heidelberg and New York, 1–18.

IPCC 1992: *Climate change 1992: the supplementary report to the IPCC scientific assessment*, ed. Houghton, J.T., Callander, B.A. and Varney, S.K. Cambridge University Press, Cambridge, 200.

IPCC 1994: *Climate change 1994: the Intergovernmental Panel on Climate Change report on the radiative forcing of climate*, ed. Houghton, J.T., Callander, B.A. and Varney, S.K. Cambridge University Press, Cambridge.

IPCC 1996: *Climate change: the IPCC second assessment report*, 3 volumes. Cambridge University Press, Cambridge.

IPCC 1998: *The regional impacts of climate change.* Cambridge University Press, Cambridge.

IUC 1997: International Unit on Conventions, Fact Sheets. United Nations Environment Programme, Geneva.

Ives, J.D. 1992: Preface. In Stone, P. (ed.), *The state of the world's mountains.* Zed Books, London, xiii–xvi.

Ives, J.D. and Messerli, B. 1989: *The Himalayan Dilemma.* Routledge, London.

Izmailova, N.N. 1977: *Wasserhaushalt kryophiler Polsterpflanzen im östlichen Pamir, Russland: Ekologia.* Academy of Sciences of the USSR **2**, 17–22.

Jacqmin, D. and Lindzen, R.S. 1985: The causation and sensitivity of the northern winter planetary waves. *Journal of Atmospheric Science* **42**, 724–745.

Jaeger, J. and Kellogg, W.W. 1983: Anomalies in temperature and rainfall during warm Arctic seasons. *Climatic Change* 34–60.

Jeffers, J.N.R. 1978: *An introduction to systems analysis with ecological applications.* Arnold, London.

Jenny, H. 1941: *Factors of soil formation.* McGraw-Hill, New York.

Jochimsen, M. and Kirchgässner, G. (eds) 1995: *Schweizer Umweltpolitik im internationalen Kontext.* Birkhäuser Verlag, Basel.

John, T.J., Samuel, R., Balraj, V. and John, R. 1998: Disease surveillance at district level. *Lancet* **352**, 58–61.

Johnson, E.A. and Larsen, C.P.S. 1991: Climatically induced change in fire frequency in the southern Canadian Rockies. *Ecology* **72**, 194–201.

Jones, P.D. and Wigley, T.M.L. 1990: Global warming trends. *Scientific American* **263**, 84–91.

Jones, R.G., Murphy, J.M. and Noger, M. 1995: Simulation of climate change over Europe using a nested regional-climate model. I: Assessment of control climate, including sensitivity to location of lateral boundaries. *Quarterly Journal of the Royal Meteorological Society* **121**, 1413–1450.

Jørgenson, S.E., Halling-Sørensen, B. and Nielson, S.N. 1996: *Handbook of environmental and ecological modeling.* CRC Lewis, London.

Kanamitsu, M. and Juang, H.-M.H. 1994: Simulation and analysis of an Indian monsoon by the NMC nested regional spectral model. *Proceedings of the International Conference on Monsoon Variability and Prediction, May 1994, Trieste, Italy.*

Karl, T.R., Jones, P.D., Knight, R.W. *et al.* 1993: Asymmetric trends of daily maximum and minimum temperature. *Bulletin of the American Meteorological Society* **74**, 1007–1023.

Karl, T.R., Wang, W.C., Schlesinger, M.E., Knight, R.W. and Portman, D. 1990: A method relating General Circulation Model simulated climate to the observed local climate, Part I: Seasonal statistics. *Journal of Climate* **2**, 1053–1079.

Karlèn, W. 1976: Lacustrine sediments and tree-limit variations as indicators of Holocene climatic fluctuations in Lappland, northern Sweden. *Geografiska Annaler* **58A**, 1–34.

Karlèn, W. 1988: Scandinavian glacial and climatic fluctuations during the Holocene. *Quaternary Science Review* **7**, 199–209.

Karlèn, W. 1993: Glaciological, sedimentological and palaeobotanical data indicating Holocene climatic change in northern Fennoscandia. In Frenzel, B., Eronen, M., Vorren, K.-D. and Gläser, B. (eds), *Oscillations of the alpine and polar tree limits in the Holocene.* Gustav Fischer Verlag, Stuttgart, 69–83.

Kassas, M. 1995: Desertification: a general review. *Journal of Arid Environments* **30**, 115–128.

Kates, R.W., Ausubel, J.H. and Berberian, M. (eds) 1985: *Climate impact assessment, SCOPE 27.* Wiley, Chichester.

Katz, R.W. and Brown, B.G. 1992: Extreme events in a changing climate: variability is more important than averages. *Climatic Change* **21**, 289–302.

Kay, B.H., Keeling, C.D. and McDonald, G. 1989: Rearing temperature influences flavivirus vector competence of mosquitoes. *Medical and Veterinary Entomology* **3**, 415–422.

Keigwin, L.D. 1996: The Little Ice Age and Medieval Warm Period in the Sargasso Sea. *Nature* **391**, 121–122.

Kharin, N.G. 1995: Change of biodiversity in ecosystems of central Asia under the impacts of desertification. In Bie, S.W. and Imevbore, A.M.A. (eds), *Biological diversity in the drylands of the world.* United Nations, New York, 23–31.

Kienast, F. 1991: Simulated effects of increasing atmospheric CO_2 and changing climate on the successional characteristics of Alpine forest ecosystems, *Landscape Ecology* **5**, 225–235.

Kienast, F. and Kuhn, N. 1989: Simulating forest succession along ecological gradients in southern Central Europe. *Vegetatio* **79**, 7–20.

Kienast, F., Wildi, O., Brzeziecki, B., Zimmermann, N. and Lemm, R. 1998: *Klimaänderung und mögliche langfristige Auswirkungen auf die Vegetation der Schweiz.* VdF Hochschulverlag, Zürich.

Kim, J.W., Chang, J.T., Baker, N.L., Wilks, D.S. and Gates, W.L. 1984: The statistical problem of climate inversion: determination of the relationship between local and large-scale climate. *Monthly Weather Review* **112**, 2069–2077.

King, G.A. and Neilson, R.P. 1992: The transient response of vegetation to climate change: a potential source of CO_2 to the atmosphere. *Water, Air and Soil Pollution* **64**, 365–383.

Kirchofer, W. (ed.) 1982: *Klimaatlas der Schweiz.* Schweizer Meteorologische Anstalt, Zürich.

Kitayama, K. 1996: Climate of the summit region of Mount Kinabalu (Borneo) in 1992, an El Niño year. *Mountain Research and Development* **16**, 65–75.

Klemes, V. 1990: Foreword. In Molnar, L. (ed.), *Hydrology of mountainous areas.* IAHS Publication 190, IAHS, Lausanne.

Klötzli, F. 1984: Neuere Erkenntnisse zur Buchengrenze in Mitteleuropa, Fukarek. *Akad. Nauka um jetn. bosne Herc.*, rad. 72, Odj. Prir. Mat. Nauka 21, P. Festschr (ed.), Sarajevo, 381–395.

Klötzli, F. 1991: Longevity and stress. In Esser, G. and Overdiek, D. (eds), *Modern ecology: basic and applied aspects.* Elsevier, Amsterdam, 97–110.

Klötzli, F. 1994: Vegetation als Spielball naturgegebener Bauherren. *Phytoecologia* **24**, 667–675.

Köppen, W.P. 1931: *Grundriss der Klimakunde.* Walter de Gruyter Verlag, Berlin.

Köppen, W.P. 1936: Das geographische System der Klimate. In Köppen, W. and Geiger, R. (eds), *Handbuch der Klimatologie I(C).* Gebrüder Borntraeger, Berlin.

Körner, C. 1993: Scaling from species to vegetation: the usefulness of functional groups. In Schulze, E.D. and Mooney, H.A. (eds), *Biodiversity and ecosystem function.* Springer-Verlag, Berlin and Heidelberg, 117–140.

Körner, C. 1994: Impact of atmospheric changes on high mountain vegetation. In Beniston, M. (ed.), *Mountain environments in changing climates.* Routledge, London, 155–166.

Körner, C. 1995: Impact of atmospheric changes on Alpine vegetation: the ecophysiological perspective. In Guisan, A., Holten, J., Spichiger, R. and Tessier, L. (eds), *Potential impacts of climate change on ecosystems in the Alps and Fennoscandian mountains.* Annex Report to the IPCC Working Group II Second Assessment Report, Publication Series of the Geneva Conservatory and Botanical Gardens, University of Geneva, 113–120.

Körner, C. 1998: Worldwide positions of Alpine treelines and their causes. In Beniston, M. and Innes, J.I. (eds), *Impacts of past, present and future climatic variability and extremes on forests.* Springer-Verlag, Heidelberg and New York, 221–230.

Körner, C., Bannister, P. and Mark, A.F. 1986: Altitudinal variation in stomatal conductance, nitrogen content and leaf anatomy in different plant life forms in New Zealand. *Oecologia* **69**, 577–588.

Körner, C. and Cochrane, P. 1983: Influence of plant physiognomy on leaf temperature on clear midsummer days in the Snowy Mountains, south-eastern Australia. *Acta Oecologica* **4**, 117–124.

Körner, C. and Diemer, M. 1987: *In situ* photosynthetic responses to light, temperature and carbon dioxide in herbaceous plants from low and high altitude. *Functional Ecology* 1, 179–194.

Körner, C., Diemer, M., Schäppi, B. and Zimmermann, L. 1995: The response of Alpine vegetation to elevated CO_2. In Koch, G.W. and Mooney, H.A. (eds), *Terrestrial ecosystem response to elevated CO_2*. Academic Press, New York, 177–196.

Körner, C. and Larcher, W. 1988: Plant life in cold climates. In Long, S.F. and Woodward, F.I. (eds), *Plants and temperature*. Company of Biologists, Cambridge, 25–57.

Körner, C. and Pelaez Menendez-Riedl, S. 1989: The significance of developmental aspects in plant growth analysis. In Lambers, H. (ed.), *Causes and consequences of variation in growth rate and productivity of higher plants*. SPB Academic Publishers, The Hague, 141–157.

Körner, C. and Renhardt, U. 1987: Dry matter partitioning and root length/leaf area ratios in herbaceous perennial plants with diverse altitudinal distribution. *Oecologia* 74, 411–418.

Kräuchi, N. 1994: *Modeling forest succession as influenced by a changing environment*. Scientific Report 69 of the Swiss Federal Institute on Forest, Snow and Landscape, Birmensdorf, Switzerland.

Kräuchi, N. and Kienast, F. 1993: Modelling subalpine forest dynamics as influenced by a changing environment, *Water, Air and Soil Pollution* 68, 185–197.

Krippendorf, J. 1984: The capital of tourism in danger. In Brugger, E.A., Messerli, B. and Bruhn, M. (eds), *The transformation of Swiss mountain regions*. Haupt Publishers, Bern, 427–450.

Kuhn, M. 1989: *The effects of long term warming on alpine snow and ice. Landscape ecological impact of climate change on alpine regions with emphasis on the Alps*. Agricultural University of Wageningen, Netherlands, 10–20.

Kuhn, M. 1993: Possible future contribution to sea-level change from small glacier. In Warrick, R.A., Barrow, E.M. and Wigley, T.M.L. (eds), *Climate and sea-level change: observations, projections and implications*. Cambridge University Press, Cambridge, 134–143.

Kullman, L. 1989: Tree-limit history during the Holocene in the Scandes mountains, Sweden, inferred from subfossil wood. *Review of Palaeobotany and Palynology* 58, 163–171.

Kullman, L. 1993: Dynamism of the altitudinal margin of the boreal forest in Sweden. In Frenzel, B., Eronen, M., Vorren, K.-D. and Gläser, B. (eds), *Oscillations of the alpine and polar tree limits in the Holocene*. Gustav Fischer Verlag, Stuttgart, 41–55.

Kutzbach, J.E. 1967: Empirical eigenvectors of sea-level pressure, surface temperature and precipitation complexes over North America. *Journal of Applied Meteorology* 133, 791–802.

Kvamme, M. 1993: Holocene forest limit fluctuations and glacier development in the mountains of southern Norway and their relevance to climate history. In Frenzel, B., Eronen, M., Vorren, K.-D. and Gläser, B. (eds), *Oscillations of the alpine and polar tree limits in the Holocene*. Gustav Fischer Verlag, Stuttgart, 99–113.

Lal, M., Jacob, D., Podzun, R. and Cubasch, U. 1995: Summer monsoon climatology simulated with a regional climate model nested in a general circulation model. *Proceedings of the International Workshop on Limited-Area and Variable-Resolution Models*, October 1995, Beijing.

LaMarche, V.C., Jr. 1973: Holocene climatic variations inferred from treeline fluctuations in the White Mountains, California. *Quaternary Research* 3, 632–660.

LaMarche, V.D., Jr., Graybill, D.A., Fritts, H.C. and Rose, M.R. 1984: Increasing atmospheric carbon dioxide: tree-ring evidence for growth enhancement in natural vegetation. *Science* 225, 1019–1021.

Lamothe, M. and D. Périard 1988: Implications of climate change for downhill skiing in Quebec. *Climate Change Digest*, Atmospheric Environment Service, Downsview, 88–103.

Lauscher, F. 1977: Ergebnisse der Beobachtungen an den nordchilenischen Hochgebirgsstationen Collahuasi und Chuquicamata. *Jahresbericht des Sonnblickvereines für die Jahre 1976–1977*, 43–67.

Le Houerou, H.N. 1989: *The grazing land ecosystems of the African Sahel*. Springer-Verlag, New York.

Leavesley, G.H. 1994: Modeling the effects of climate change on water resources: a review. In Frederick, K.D. and Rosenberg, N. (eds), *Assessing the impacts of climate change on natural resource systems*. Kluwer, Dordrecht, 179–208.

Leonard, E. 1986: Use of lacustrine sedimentary sequences as indicators of Holocene glacier history, Banff National Park, Alberta, Canada. *Quaternary Research* 26, 218–231.

Leung, R.L. and Ghan, S.J. 1995: A subgrid parameterization of orographic precipitation. *Theoretical and Applied Climatology* 51, 175–189.

Levin, S.A. 1992: The problem of pattern and scale in ecology. *Ecology* 73, 1943–1967.

Lins, H., Michaels, P.J. and Leavesley, G.H. 1990: Hydrology and water resources. In Tegart, W.J.McG., Sheldon, G.W. and Griffiths, D.C. (eds), *Climate change: the IPCC impacts assessment*. Australian Government Publishing Service, Canberra, Chapter 4.

Liu, Y.F., Giorgi, F. and Washington, W.M. 1994: Simulation of summer monsoon climate over eat Asia with an NCAR regional climate model. *Monthly Weather Reviews* 122, 2331–2348.

Liverman, D.M. and O'Brien, K.L. 1991: Global warming and climate change in Mexico. *Global Environmental Change: Human and Policy Dimensions* 1, 351–364.

Loevisohn, M. 1994: Climatic warning and increased malaria incidence in Rwanda. *Lancet* 343, 714–718.

Lorenz, E.N. 1963: Deterministic non-periodic flow. *Journal of Atmospheric Science* 20, 130–141.

Lorenz, E.N. 1968: Climate determinism. *Meteorological Monographs* 8. AMS Publications, Boston.

Lorenz, E.N. 1982: Atmospheric predictability experiments with a large numerical model. *Tellus* 34, 505–513.

Lough, J.M., Wigley, T.M.L. and Palutikof, J.P. 1984: Climate and climate impact scenarios for Europe in a warmer world. *Journal of Climatology and Applied Meteorology* 23, 1673–1684.

Lubchenco, J., Olsen, A.M., Brubaker, L.B. *et al.* 1991: The sustainable biosphere initiative. An ecological research initiative. *Ecology* 72, 371–412.

Luckman, B.H. 1994: Using multiple high-resolution proxy climate records to reconstruct natural climate variability: an example from the Canadian Rockies. In Beniston, M. (ed.), *Mountain environments in changing climates*. Routledge, London, 42–59.

Lundgren, L. 1980: Comparison of surface runoff and soil loss from runoff plots in forest and small scale agriculture in the Usambara Mountains, Tanzania. University of Stockholm, Department of Physical Geography.

Lynch, P., McBoyle, G.R. and Wall, G. 1981: A ski season without snow. In Phillips, D.W. and McKay, G.A. (eds), *Canadian Climate in Review B 1980.* Atmospheric Environment Service, Toronto, 42–50.

MacArthur, R.H. 1972: *Geographical ecology.* Harper and Row, New York.

Machenhauer, B., Jacob, D. and Bozert, M. 1994: Using the MPI's nested limited area model for regional climate simulations. WMO/TO-592, 758–760.

Mackay, J.R. 1990: Seasonal growth bands in pingo ice. *Canadian Journal of Earth Science* **27**, 1115–1125.

Mackay, J.R. 1992: Frequency of ice-wedge cracking 1967–1987 at Garry Island, Western Arctic Coast, Canada. *Canadian Journal of Earth Science* **29**, 11–17.

Magny, M. 1995: Successive oceanic and solar forcing indicated by Younger Dryas and early Holocene climatic oscillations in the Jura. *Quaternary Research* **43**, 279–285.

Maisch, M. 1992: *Die Gletcher Graubündens: Rekonstruktion und Auswertung der Gletscher und deren Veränderungen seit dem Hochstand von 1850 im Gebiet der östlichen Schweizer Alpen (Bündnerland und angrenzende Regionen).* Publication Series of the Department of Geography of the University of Zürich.

Manabe, S. and Terpstra, T.B. 1974: The effects of mountains on the general circulation of the atmosphere as identified by numerical experiments. *Journal of Atmospheric Science* **31**, 3–42.

Mandelbrot, B.B. 1983: *The fractal geometry of nature.* W.H. Freeman, San Francisco.

Marinucci, M.R., Giorgi, F., Beniston, M., Wild, M., Tschuck, P. and Bernasconi, A. 1995: High resolution simulations of January and July climate over the Western Alpine region with a nested regional modeling system. *Theoretical and Applied Climatology* **51**, 119–138.

Markgraf, V. 1980: Pollen dispersal in a mountain area. *Grana* **19**, 127–146.

Markham, A. and Malcolm, J. 1996: Biodiversity and wildlife conservation: adaptation to climate change. In Smith, J., Bhatti, N., Menzhulin, G., Benioff, R., Campos, M., Jallow, B. and Rijsberman, F. (eds), *Adaptation to climate change: assessment and issues.* Springer-Verlag, New York, 384–401.

Martens, W.J.M., Niessen, L.W., Rotmans, T.H. and McMichael, A.J. 1995: Potential impact of global climate change on malaria risk. *Environmental Health Perspectives* **103**, 458–464.

Martin, E. 1995: Modélisation de la climatologie nivale des Alpes françaises. PhD Dissertation, Université Paul Sabatier, Toulouse, France.

Martin, E. and Durand, Y. 1998: Precipitation and snow cover variability in the French Alps. In Beniston, M. and Innes, J.L. (eds.), *The impacts of climate change on forests.* Springer-Verlag, Heidelberg and New York, 81–92.

Martin, P. and Lefebvre, M. 1995: Malaria and climate: sensitivity of potential transmission to climate. *Ambio* **24**, 200–207.

Martinec, J. and Rango, A. 1989: Effects of climate change on snow melt runoff patterns. *Remote sensing and large-scale global processes, Proceedings of the IAHS Third International Assembly,* Baltimore, MD, May 1989, IAHS Publ. 186, 31–38.

Matthes, F.E. 1939: Report of the Committee on Glaciers, April 1939. *Transactions of the American Geophysical Union* **20**, 518–523.

Matthews, J. 1991: The late Neoglacial ('Little Ice Age') glacier maximum in southern Norway: new ^{14}C dating evidence and climatic implications. *The Holocene* **1**, 219–233.

Matthews, J. 1993: Deposits indicative of Holocene climatic fluctuations in the timberline areas of Northern Europe: some physical proxy data sources and research aproaches. In Frenzel, B., Eronen, M., Vorren, K-D. and Gläser, B. (eds), *Oscillations of the alpine and polar tree limits in the Holocene.* Gustav Fischer Verlag, Stuttgart, 85–97.

Maunder, W.J. 1986: *The uncertainty business.* Methuen, London.

Mayer-Tasch, P.C., Molt, W. and Tiefenthaler, H. (eds), 1990: *Transit: Das Drama der Mobilität: Wege zu einer humanen Verkehrspolitik.* Schweizer Verlagshaus, Zürich.

McBoyle, G.R. and Wall, G. 1987: The impact of CO_2-induced warming on downhill skiing in the Laurentians. *Cahiers de Géographie de Québec* **31**, 39–50.

McGregor, J.L. and Walsh, K. 1993: Nested simulations of perpetual January climate over the Australian region. *Journal of Geophysical Research* **98**, 23283–23290.

McGregor, J.L. and Walsh, K. 1994: Climate change simulations of Tasmanian precipitation using multiple nesting. *Journal of Geophysical Research* **99**, 20889–20905.

McMurtrie, R.E. and Comins, H.N. 1996: The temporal response of forest ecosystems to doubled atmospheric CO_2 concentration. *Global Change Biology* **2**, 49–57.

McNeely, J.A. 1990: Climate change and biological diversity: policy implications. In Boer, M.M. and de Groot, R.S. (eds), *Landscape-ecological impact of climatic change.* IOS Press, Amsterdam.

Mearns, L.O., Giorgi, F., McDaniel, L. and Brodeur C.S. 1994: Analysis of daily variability of precipitation in a nested regional climate model: comparison with observations and doubled CO_2 results. *Global and Planetary Change* **27**, 17–27.

Mehta, M. 1995: Cultural diversity in the mountains: issues of integration and marginality in sustainable development. *Conference on the Mountain Agenda,* Lima, Peru, February 1995.

Melillo, J.M., Prentice, I.C., Farquhar, G.D., Schulze, E.-D. and Sala, O.E. 1996: Terrestrial biotic responses to environmental change and feedbacks to climate. In Houghton, J.T., Meira Filho, L.G., Callander, B.A., Harris, N., Kattenberg, A. and Maskell, K. (eds), *Climate change 1995. The science of climate change.* Cambridge University Press, Cambridge, 447–481.

Messerli, B. 1973: Problems of vertical and horizontal arrangement in the high mountains of the extreme arid zone. *Arctic and Alpine Research* **5**, 139–148.

Messerli, P. 1989: *Mensch und Natur im alpinen Lebensraum: Risiken, Chancen, Perspektiven.* Haupt-Verlag, Bern and Stuttgart.

Michaelson, J. and Thompson, L.G. 1992: A comparison of proxy records of El Niño/Southern Oscillation. In, Diaz, H.F. and Markgraf, V. (eds), *El Niño, historical and paleoecological aspects of the Southern Oscillation.* Cambridge University Press, Cambridge, 323–348.

Miller, A.J., Cayan, D.R., Barnett, T.P., Graham, N.E. and Oberhuber, J.M. 1994: Interdecadal variability of the Pacific Ocean: model response to observed heat flux and wind stress anomalies. *Climate Dynamics* **9**, 287–301.

Miller, J T. Jr. 1996: *Living in the environment.* Wadsworth, Belmont, CA.

Mirza, M.Q. 1997: The runoff sensitivity of the Ganges river basin to climate change and its implications. *Journal of Environmental Hydrology* **5**, 1–13.

Mirza, M.Q. and Dixit, A. 1997: Climate change and water management in the GBM basins. *Water Nepal* **5**, 71–100.

Monasterio, M. (ed.) 1980: *Estudios ecológicos en los páramos andinos.* Ediciones, Universidad de los Andes, Mérida, Venezuela.

Nash, L.L. and Gleick, P.H. 1991: Sensitivity of streamflow in the Colorado Basin to climatic changes. *Hydrology* **125**, 221–241.

Neilson, R.P. 1986: High-resolution climatic analysis and southwest biogeography. *Science* **232**, 27–34.

Neilson, R.P. 1995: A model for predicting continental-scale vegetation distribution and water balance. *Ecological Application* **5**, 362–385.

Neilson, R.P., King, G.A. and Koerper, G. 1992: Toward a rule-based biome model. *Landscape Ecology* **7**, 27–43.

Nesje, A., Dahl, S.O., Løvlie, R. and Sulebak, J.R. 1994: Holocene glacier activity at the southwestern part of Hardangerjøkulen, central-south Norway: evidence from lacustrine sediments. *The Holocene* **4**, 377–382.

Nesje, A. and Johannessen, T. 1992: What were the primary forcing mechanisms of high-frequency Holocene climate and glacier variations? *The Holocene* **2**, 79–84.

Nesje, A., Kvamme, M., Rye, N. and Løvlie, R. 1991: Holocene glacier and climate history of the Jostedalsbreen region, western Norway: evidence from lake sediments and terrestrial deposits. *Quaternary Science Reviews* **10**, 87–114.

New Zealand Ministry for the Environment 1990: *Climatic change: a review of impacts on New Zealand.* DSIR Publishing, Wellington.

Nicolis, G. and Prigogine, I. 1989: *Exploring complexity.* W.H. Freeman, San Francisco.

Nigam, S., Held, I.M. and Lyons, S.W. 1988: Linear simulations of the stationary eddies in a GCM. Part II: The mountain model. *Journal of Atmospheric Science* **45**, 1433–1452.

Njiro, E.I. 1998: *Montane ecosystems: characteristics and conservation.* Mountain Forum On-Line Library.

Noble, I. and Gitay, H. 1998: Climate change in desert regions. In Watson, R.T., Zinyowera, M. and Moss, R. (eds), IPCC 1998, *The regional impacts of climate change.* Cambridge University Press, pp. 191–217.

Nordhaus, W.D. 1994: *Managing the global commons: the economics of climate change.* MIT Press, Cambridge, MA.

Nordhaus, W.D. and Yang, Z. 1996: A regional and dynamic general equilibrium model of alternative climate change strategies. *American Economic Review* **86**, 741–762.

Oberbauer, S.F. and Billings, W.D. 1981: Drought tolerance and water use by plants along an alpine topographic gradient. *Oecologia* **50**, 325–331.

Obrebska-Starkel, B. 1990: Recent studies on Carpathian meteorology and climatology. *International Journal of Climatology* **10**, 79–88.

Oerlemans, J. (ed.) 1989: *Glacier fluctuations and climate change.* Kluwer, Dordrecht.

Oerlemans, J. 1993: Modelling of glacier mass balance. In Peltier, W.R. (ed.), *Ice in the climate system.* NATO ASI Series I, 12. Springer-Verlag, Berlin, 101–116.

Oerlemans, J. 1994: Quantifying global warming from the retreat of glaciers. *Science* **264**, 243–245.

Oerlemans, J. and Fortuin, J.P.F. 1992: Sensitivity of glaciers and small ice caps to greenhouse warming. *Science* **258**, 115–117.

Oerlemans, J. and Hoogendorn, N.C. 1990: Mass-balance gradients and climatic changes. *Journal of Glaciology* **35**, 399–405.

Oeschger, H. and Langway, C.C., Jr (eds) 1989: *The environmental record in glaciers and ice sheets.* Wiley, New York.

Ogden, J. 1985: An introduction to plant demography with special reference to New Zealand trees. *New Zealand Journal of Botany* **23**, 751–772.

Ohlendorf, C., Niessen, F. and Weissert, H. 1997: Glacial varve thickness and 127 years of instrumental climate data: a comparison. *Climatic Change* **36**, 391–411.

Ollier, C.D. 1976: Catenas in different climates. In Derbyshire, E. (ed.), *Geomorphology and climate.* Wiley, London, 137–170.

O'Neill, R.V. 1988: Hierarchy theory and global change. In Rosswall, T., Woodmansee, R.G. and Risser, P.G. (eds), *Scales and global change.* Wiley, London.

Oort, A.H. and Liu, H. 1993: Upper-air temperature trends over the globe 1958–1989. *Journal of Climatology* **6**, 292–307.

Orlanski, J. 1975: A rational subdivision of scales for atmospheric processes. *Bulletin of the American Meteorological Society* **56**, 527–530.

OTA (US Congress, Office of Technology Assessment) 1993: *Preparing for an uncertain climate*, 2 vols, OTA-O-567 and 568. US Government Printing Office, October, Washington, DC.

Overpeck, J.T., Rind, D. and Goldberg, R. 1990: Climate-induced changes in forest disturbance and vegetation. *Nature* **343**, 51–53.

Ozenda, P. 1985: *La végétation de la chaîne alpine dans l'espace montagnard européen.* Masson, Paris.

Ozenda, P. and Borel, J.-L. 1991: *Les conséquences écologiques possibles des changements climatiques dans l'arc alpin.* Rapport Futuralp 1, International Centre for Alpine Environment (ICALP), Le Bourget-du-lac, France.

Parry, M.L. 1978: *Climatic change, agriculture and settlement.* Dawson-Archon Books, Folkestone, UK.

Parry, M.L., Carter, T.R. amd Konijn, N.T. 1990: Agriculture and forestry. In Tegart, W.J.McG., Sheldon, G.W. and Griffiths, D.C. (eds), *Climate change: the IPCC impacts assessment.* Australian Government Publishing Service, Canberra, ch 2.

Parry, M.L., de Rozari, M.B., Chong, A.L. and Panich, S. (eds) 1992: *The potential socio-economic effects of climate change in South-East Asia.* Nairobi: UN Environmental Programme.

Parsons, R.B. 1978: Soil–geomorphology relationships in mountains of Oregon. *Geoderma* **21**, 25–39.

Peine, J.D. and Fox, D.G. 1995: Wilderness environmental monitoring: assessment and atmospheric effects. In Cordell, K. (ed.), *The Status of Wilderness Research.* General Technical Report, GTR-SE-332, USDA-Forest Service, Asheville, 91.

Peine, J.D. and Martinka, C.J. 1992: Effects of climate change on mountain protected areas: implications for management. *Proceedings of the Fourth World Congress on National Parks and Protected Areas, Caracas, Venezuela*, 10–21 February 1992.

Penck, A. and Bruckner, E. 1909: *Die Alpen im Eiszeitaler.* Tauchnitz Verlag, Leipzig.

Perry, A.H. 1971: Climatic influences on the Scottish ski industry. *Scottish Geographical Magazine* **87**, 197–201.

Perry, A.H. forthcoming: Impacts of climate change on tourism, energy and transportation. In *IPCC Third Assessment Report, Impacts of Climate Change on Europe.* Cambridge University Press, Cambridge (in preparation).

Perry, A.H. and Smith, K. 1996: *Recreation and tourism.* Climate Change Impacts Report, UK Department of the Environment, London.

Peters, R.L. and Darling, J.D.S. 1985: The greenhouse effect and nature reserves: global warming would diminish biological diversity by causing extinctions among reserve species. *Bioscience* **35**, 707–717.

Peterson, D.L. 1994: Recent changes in the growth and establishment of subalpine conifers in western North America, In Beniston, M. (ed.), *Mountain environments in changing climates.* Routledge, London, 234–243.

Peterson, D.L., Arbaugh, M.J., Robinson, L.J. and Derderian, B.R. 1990: Growth trends of whitebark pine and lodgepole pine in a subalpine Sierra Nevada forest, California, U.S.A. *Arctic and Alpine Research* **22**, 233–243.

Pfister, C. 1985a: *Klimageschiche der Schweiz, 1525–1860,* vol. 1, Academia Helvetica 6. P. Haupt, Bern.

Pfister, C. 1985b: Snow cover snowlines and glaciers in central Europe since the 16th century. In Tooley, M.J. and Sheail, G.M. (eds), *The climatic scene.* Allen & Unwin, London, 154–174.

Pfister, C. 1994: Climate in Europe during the Late Maunder Minimum period. In Beniston, M. (ed.), *Mountain environments in changing climates.* Routledge, London, 60–90.

Pils, M., Glauser, P. and Siegrist, D. (eds) 1998: *Green Paper for the Alps.* European Union Directorate XII Publication and Nature Friends for the Alps, Vienna.

Pisek, A. and Cartellieri, E. 1941: Der Wasserverbrauch einiger Pflanzenvereine. *Zeitschrift für Berliner Wissenschaftlich Botanik* **90**, 282–291.

Pittock, A.B. and Salinger, M.J. 1982: Towards regional scenarios for a CO_2-warmed earth. *Climatic Change* **21**, 23–40.

Podzun, R., Cress, A., Majewski, D. and Renner, V. 1995: Simulation of European climate with a limited area model. Part II: AGCM boundary conditions. *Contributions to Atmospheric Physics* **72**, 53–70.

Porter, S.C. 1979: Hawaiian glacial ages. *Quaternary Research* **12**, 161–187.

Porter, S.C. and Denton, G. 1967: Chronology of neoglaciation in the North American Cordillera. *American Journal of Science* **265**, 177–210.

Prentice, I.C. 1986: Vegetation response to past climate variation. *Vegetatio* **67**, 131–141.

Prentice, I.C. 1992: Climate and long-term vegetation dynamics. In Glenn-Lewin, D.C., Peet, R.A., Veblen, T.T. (eds), *Plant succession: theory and prediction.* Chapman & Hall, London, 293–339.

Prentice, I.C., Cramer, W., Harrison, S.P., Leemans, R., Monserud, R.A. and Solomon, A.M. 1992: A global biome model based on plant physiology and dominance, soil properties and climat. *Journal of Biogeography* **19**, 117–134.

Price, L.W. 1981: *Mountains and man.* University of California Press, Berkeley.

Price, M.F. 1990: Temperate mountain forests: common-pool resources with changing, multiple outputs for changing communities. *Natural Resources Journal* **30**, 685–707.

Price, M.F. 1993: Patterns of the development of tourism in mountain communities. In Allan, N.J.R. (ed.), *Mountains at risk: current issues in environmental studies.* Kluwer, Dordrecht.

Price, M.F. 1994: Should mountain communities be concerned about climate change? In Beniston, M. (ed.), *Mountain environments in changing climates.* Routledge, London, 431–451.

Prinns, E.M., Menzel, W.P. and Feltz, J.M. 1998: Using geostationary meteorological satellites to monitor trends in biomass burning: a four-year case study in South America. *Abstracts of the International Workshop on Biomass Burining and its Inter-relationships with the Climate System. Wengen, Switzerland, September 1998.*

Quezel, P. and Barbero, M. 1990: Les forêts méditerranéennes: problèmes posés par leur signification historique, écologique et leur conservation. *Acta Botanica Malacitana* **15**, 145–178.

Raich, J.W., Rastetter, E.B., Melillo, J.M. *et al.* 1991: Potential net primary production in South America. *Ecological Applications* **1**, 399–429.

Rameau, J.C., Mansion, D., Dumé, G., Lecointe, A., Timbal, J., Dupont, P. and Keller, R. 1993: *Flore forestière française, guide ecologique illustré.* Lavoisier TEC and DOC Diffusion, Paris.

Rapp, A. 1960: Recent developments on mountain slopes in Kärkevagge and surroundings. *Geografiska Annaler* **42**, 1–158.

Rebetez, M. and Beniston, M. 1998: Changes in temperature variability in relation to shifts in mean temperatures in the Swiss Alpine region this century. In Beniston, M. and Innes, J.L. (eds), *The impacts of climate change on forests.* Springer-Verlag, Heidelberg and New York, 55–67.

Rebetez, M., Lugon, R. and Baeriswyl, P.A. 1997: Climatic change and debris flows in high mountain regions. *Climatic Change* **36**, 371–389.

Richard, J.L. 1985: Pelouses xerophiles alpines des environs de Zermatt (Valais, Suisse). *Botanica Helvetica* **95**, 193–211.

Richter, L.K. 1989: *The politics of tourism in Asia.* University of Hawaii Press, Honolulu.

Riebsame, W.E. 1989: *Assessing the social implications of climate fluctuations.* United Nations Environment Programme, Nairobi.

Riebsame, W.E., Woodmansee, R. and Peters, N. 1995: Complex river basins. In Strzepek, K.M. and Smith, J.B. (eds), *As climate changes: international impacts and implications.* Cambridge University Press, Cambridge, 57–91.

Rind, D. 1988: Dependence of warm and cold climate depiction on climate model resolution. *Journal of Climate* **1**: 965–997.

Rind, D. and Overpeck, J. 1993: Hypothesized causes of decadal-to-century climate variability: climate model results. *Quaternary Science Reviews* **12**, 357–374.

Rind, D. and Peteet, D. 1985: Terrestrial conditions at the last glacial maximum and CLIMAP sea-surface temperature estimates: Are they consistent? *Quaternary Research* **24**, 1–22.

Rizzo, B. and Wiken, E. 1992: Assessing the sensitivity of Canada's ecosystems to climatic change. *Climate Change* **21**, 37–54.

Robin, G. de Q. 1983: *The climatic record in polar ice sheets.* Cambridge University Press, Cambridge.

Rochefort, R.R., Little, R.L., Woodward, A. and Peterson, D.L. 1994: Changes in sub-alpine tree distribution in western North America: a review of climatic and other causal factors. *The Holocene* **4**, 89–100.

Rodda, J. C. 1994: Mountains: a hydrological paradox or paradise? In Keller, H.M. (ed.), *Hydrologie Kleiner Einzugsgebiete, Gedenkschrift: Beitrage zur Hydrologie der Schweiz* **35**, 41–51.

Romme, W.H. and Turner, M.G. 1991: Implications of global climate change for biogeographic patterns in the greater Yellowstone ecosystem. *Conservation Biology* **5**, 373–386.

Rongers, W.A. 1993: The conservation of forest resources of Eastern Africa: past influences, present practices and future needs In Lovett, J.C. and Wasser, S.K. (eds), *Biogeography and ecology of the rain forests of eastern Africa.* Cambridge University Press, Cambridge, 222–247.

Rosenzweig, C., Parry, M.L. and Fischer, G. 1993: *Climate change and world food supply.* Oxford University Press, Oxford.

Rotach, M., Wild, M., Tschuck, P., Beniston, M. and Marinucci, M.R. 1996: A double CO_2 experiment over the Alpine region with a nested GCM–LAM modeling approach. *Theoretical and Applied Climatology* **57**, 209–227.

Röthlisberger, F. 1986: *10,000 Jahre Gletschergeschichte der Erde.* Verlag Sauerlander, Aarau and Frankfurt.

Rupke, J. and Boer, M.M. (eds) 1989: Landscape Ecological Impact of Climatic Change on Alpine Regions, with Emphasis on the Alps, Discussion report prepared for European conference on landscape ecological impact of climatic change, Agricultural University of Wageningen and Universities of Utrecht and Amsterdam, Wageningen, Utrecht and Amsterdam.

Salinger, M.J., Williams, J.M. and Williams, W.M. 1989: *CO_2 and climate change: impacts on agriculture.* New Zealand Meteorological Service, Wellington.

Santer, B.D., Wigley, T.M.L., Schlesinger, M.E. and Mitchell, J.F.B. 1990: *Developing climate scenarios from equilibrium GCM results.* Max-Planck-Institut für Meteorologie Report 47. Hamburg.

Sasaki, H., Kida, H., Koide, T. and Chiba, M. 1995: The performance of long-term integrations of a limited area model with the spectral boundary coupling method. *Journal of the Meteorological Society of Japan* **73**, 273–278.

Schaeffer, M. 1998: Developing integrated transient scenarios of global climatic change. In *The Climate LINK Project Workshop Abstracts, May 1998.* Climatic Research Unit, University of East Anglia, 12–14.

Schaer, C., Frei, C., Lüthi, C. and Davies, H.C. 1996: Surrogate climate change scenarios for regional climate models. *Geophysical Research Letters* **23**, 185–209.

Schneider, T. (ed.) 1989: *Atmospheric ozone research and its policy implications.* Elsevier, Amsterdam.

Schneider, U. 1992: Die Verteilung des troposphärischen Ozons in Bayrischen Nordalpenraum. PhD Dissertation. University of Mainz.

Schreir, H. and Shah, P.B. 1996: Water dynamics and population pressure in the Nepal Himalayas. *Geojournal* **40**, 45–51.

Schubert, C. 1992: The glaciers of the Sierra Nevada de Mérida (Venezuela): a photographic comparison of recent deglaciation. *Erdkunde* **46**, 58–64.

Schweingruber, F.H. 1988: A new dendroclimatic network for western North America. *Dendrochronologia* **6**, 171–180.

Schweingruber, F.H., Bräker, O.U. and Schär, E. 1979: Dendroclimatic studies on conifers from central Europe and Great Britain. *Boreas* **8**, 427–52.

Semazzi, F.H.M., Lin, N.-H., Lin, Y.-L. and Giorgi, F. 1994: A nested modeling study of the Sahelian climate response to sea-surface temperature anomalies. *Geophysical Research Letters* **20**, 2897–2900.

Senarclens-Grancy, W. 1958: Zur Glacialgeologie der Ötztales und seine Umgebung, *Mitteilungen der Geologisches Gesellschaft Wien* **49**, 257–314.

Sharma, N.P., Damhaug, T., Gilgan-Hunt, E., Grey, D., Okaru, V. and Rothberg, D. 1996: *African water resources: challenges and opportunities for sustainable development.* World Bank Technical Paper 331. World Bank, Washington, DC.

Shiklomanov, I. 1993: World freshwater resources. In Gleick, P. (ed.), *Water in crisis: a guide to the world's freshwater resources.* Oxford University Press, Oxford, 13–24.

Shiva, V. 1988: *Staying alive: women, ecology and survival in India.* New Delhi: Kali for Women.

Shriner, D.S. and Street, R.B. 1998: North America. In Watson, R.T., Zinyowerd, M. and Moss, R. (eds), *The regional impacts of climate change.* Cambridge University Press, Cambridge, 253–330.

Shugart, H.H. 1984: *A theory of forest dynamics: the ecological implications of forest succession models.* Springer-Verlag, New York.

Shugart, H.H. and West, D.C. 1977: Development of an Appalachian deciduous forest succession model and its application to assessment of the impact of the chestnut blight. *Journal of Environmental Management* **5**, 161–179.

Sinha, S.K. and Swaminathan, M.S. 1991: Deforestation, climate change and sustainable nutrition security. *Climatic Change* **19**, 201–209.

Slaymaker, O. 1990: Climate change and erosion processes in mountain regions of western Canada. *Mountain Research and Development* **10**, 171–182.

Slayter, R.O., Cochrane, P.M. and Galloway, R.W. 1984: Duration and extent of snow cover in the Snowy Mountains and a comparison with Switzerland, *Search* **15**, 327–331.

Slayter, R.O. and Noble, I.R. 1992: Dynamics of montane treelines. In Hansen, A.J. and di Castri, F. (eds), *Landscape boundaries: consequences for biotic diversity and ecological flows.* Springer-Verlag, New York, 346–359.

SMA 1997: *Monthly weather and statistics bulletin, December 1997.* Swiss Meteorological Agency, Zürich.

Smidt, S. 1991: *Messungen nasser Freilanddepositionen der Forstlichen Bundesversuchsanstalt.* FBVA-Berichte, ISSN 1013-0713 50, Nasse Deposition, Austria.

Smith, R.B. 1979: The influence of mountains on the atmosphere, *Advances in Geophysics*, vol. 21, Academic Press, New York, 87–230.

Smith, W.K. and Knapp, A.K. 1990: Ecophysiology of high mountain forests. In Osmond, C.B., Pitelka, L.F. and Hidy, G.M. (eds), *Plant biology of the basin and range.* Springer-Verlag, New York, 87–142.

Solbrig, O. 1984: Tourism. *Mountain Research and Development* **4**, 181–185.

Solomon, A.M. 1986: Transient response of forests to CO₂-induced climate change: simulation modeling experiments in eastern North America. *Oecologia* **68**, 567–579.

Stanners, D. and Bourdeau, P. (eds) 1995: *Europe's environment: the Dobris assessment.* European Environment Agency, Copenhagen.

Starosolszky, O. and Melder, O.M. 1989: *Hydrology of disasters.* James and James, London.

Steinhauser, F. 1970: Die säkularen Änderungen der Schneedeckenverhältnisse in Österreich. *66-67 Jahresbericht des Sonnblick-Vereines 1970–1971.* Verlag der Sonnblick-Stiftungs Vienna.

Stocks, B.J. 1993: Global warming and forest fires in Canada. *The Forestry Chronicle* **69**, 290–293.

Stone, P.B. (ed.) 1992: *The state of the world's mountains.* Zed Books, London.

Street, R.B. and Melnikov, P.I. 1990: Seasonal snow, cover, ice and permafrost. In Tegart, W.J.McG., Sheldon, G.W. and Griffiths, D.C. (eds), *Climate change: the IPCC impacts assessment.* Australian Government Publishing Service, Canberra, ch. 7.

Street, R.B. and Semenov, S.M. 1990: Natural terrestrial ecosystems. In Tegart, W.J.McG., Sheldon, G.W. and Griffiths, D.C. (eds), *Climate change: the IPCC impacts assessment report.* Australian Government Publishing Service, ch. 3.

Street-Perrott, F.A. and Perrott, R.A. 1990: Lake levels and climate reconstructions. In Hecht, A. (ed.), *Palaeo-climate analysis and modeling.* Wiley, New York, 291–340.

Swiss Federal Office of Transportation, Communications and Energy 1991: *Service d'étude des transports: les transports: hier, aujourd'hui, demain.* Publication SET 1/91.

Swiss Federal Statistical Office/Swiss Agency for the Environment, Forests and Landscape 1997: *The environment in Switzerland,* ch. 25.

Swiss Re 1997: *Climatic change and the insurance industry.* Swiss Reinsurance Company, Zürich.

Szeicz, J.M. and MacDonald, G.M. 1995: Recent white spruce dynamics at the subarctic alpine treeline of northwestern Canada. *Journal of Ecology* **83**, 873–885.

Teranes, J.L. and McKenzie, J.A. 1995: Evidence for rapid climate changes during the twentieth century from high-resolution oxygen isotope stratigraphy in chemically varved lacustrine sediments. *Terra Nova* **7**, 218.

Tessier, L., de Beaulieu, J.-L., Couteaux, M. *et al.* 1993: Holocene palaeoenvironments at the timberline in the French Alps: a multidisciplinary approach. *Boreas* **22**, 244–254.

Tessier, L., Guibal, F. and Schweingruber, F.H. 1997: Research strategies in dendroecology and dendroclimatology in mountain environments. *Climatic Change* **36**, 499–517.

Thinon, M. 1992: L'analyse pédoanthracologique: aspects méthodologiques et applications, PhD Dissertation, University of Aix-Marseille III, France.

Thompson, L.G. 1991: Ice core records with emphasis on the global record of the last 2000 years. In Bradley, R.S. (ed.), *Global changes of the past.* University Corporation for Atmospheric Research, Boulder, CO, 201–224.

Thompson, L.G. 1992: Ice core evidence from Peru and China. In Bradley, R.S. and Jones, P.D. (eds), *Climate since A.D. 1500.* Routledge, London, 517–548.

Thompson, L.G., Davis, M., Mosley-Thompson, E. and Liu, K. 1988b: Pre-Incan agricultural activity recorded in dust layers in two tropical ice cores. *Nature* **336**, 763–765.

Thompson, L.G. and Mosley-Thompson, E. 1987: Evidence of abrupt climatic change during the last 1500 years recorded in ice cores from the tropical Quelccaya Ice Cap, Peru. In Berger, W.H. and Labeyrie, L.D. (eds), *Abrupt climatic change.* Reidel, Dordrecht, 99–110.

Thompson, L.G., Mosley-Thompson, E. and Arnao, B.M. 1984: Major El Niño/Southern Oscillation events recorded in stratigraphy of the tropical Quelccaya Ice Cap. *Science* **226**, 50–52.

Thompson, L.G., Mosley-Thompson, E., Bolzan, J.F. and Koci, B.R. 1985: A 1500 year record of tropical precipitation in ice cores from the Quelccaya Ice Cap, Peru. *Science* **229**, 971–973.

Thompson, L.G., Mosley-Thompson, E., Dansgaard, W. and Grootes, P.M. 1986: The Little Ice Age as recorded in the stratigraphy of the tropical Quelccaya Ice Cap. *Science* **234**, 361–364.

Thompson, L.G., Mosley-Thompson, E., Davis, M.E. *et al.* 1989. Holocene–late Pleistocene climatic ice core records from Qinghai–Tibetan plateau. *Science* **246**, 474–477.

Thompson, L.G., Mosley-Thompson, E., Davis, M.E. *et al.* 1990: Glacial stage ice core records from the subtropical Dunde Ice Cap, China. *Annals of Glaciology* **14**, 288–298.

Thompson, L.G., Mosley-Thompson, E., Davis, M.E., Lin, N., Yao, T., Dyurgerov, M. and Dai, J. 1993: 'Recent warming': ice core evidence from tropical ice cores, with emphasis on central Asia. *Global and Planetary Change* **7**, 145–156.

Thompson, L.G., Mosley-Thompson, E., Davis, M.E. *et al.* 1995: Late glacial stage and Holocene tropical ice core records from Huascarán, Peru. *Science* **269**, 46–50.

Thompson, L.G., Mosley-Thompson, E., Wu, X. and Xie, Z. 1988a: Wisconsin/Würm glacial stage ice in the sub-tropical Dunde Ice Cap, China. *Geojournal* **17**, 517–523.

Thornley, J.H.M. and Cannell, M.G.R. 1996: Temperate forest responses to carbon dioxide, temperature and nitrogen: a model analysis. *Plant Cell and Environment* **19**, 1331–1348.

Tranquillini, W. 1979: *Physiological ecology of the Alpine timberline.* Springer-Verlag, Berlin.

Tranquillini, W. 1993: Climate and physiology of trees in the Alpine timberline regions. In Frenzel, B., Eronen, M., Vorren, K.-D. and Gläser, B. (eds), *Oscillations of the alpine and polar tree limits in the Holocene.* Gustav Fischer Verlag, Stuttgart, 127–135.

Trenberth, K.E. 1990: Recent observed interdecadal climate changes in the Northern Hemisphere. *Bulletin of the American Meteorological Society* **71**, 988–993.

Trenberth, K.E. and Hoar, T.J. 1995: The 1990–1995 El Niño–Southern Oscillation event: longest on record. *Geophysical Research Letters* **23**, 57–60.

Trenberth, K.E. and Hurrell, J.W. 1994: Decadal atmosphere–ocean variations in the Pacific. *Climate Dynamics* **9**, 303–319.

UNEP 1992: *Glaciers and the environment.* UNEP/GEMS Environment Library 9.

UNEP 1993: *Environmental data report.* Oxford and New York.

UNEP 1997: *Global environment outlook: an overview.* UNEP in collaboration with the Stockholm Environment Institute; Oxford University Press, New York and Oxford.

Urbach, F. 1989: Potential effects of altered solar ultraviolet radiation on human skin cancer. *Photochemistry and Photobiology* **50**, 507–514.

van der Hammen, T. 1984: Datos eco-climáticos de la transecta Buritaca y alrededores (Sierra Nevada de Santa Marta). In van der Hammen, T. and Ruiz, P. (eds), *La Sierra Nevada de Santa Marta (Colombia), Transecta Buritaca–La Cumbre.* J. Cramer, Berlin, 45–66.

VAW 1993: *Greenhouse gases, isotopes and trace elements in glaciers as climate evidence for the Holocene: report on the ESF/EPC workshop, Zürich, 27–28 October 1992.* VAW Arbeitsheft, Zürich.

Veblen, T.T., Ashton, D.H., Schlegel, F.M. and Veblen, A.T. 1977: Plant succession in a timberline depressed by vulcanism in south-central Chile. *Journal of Biogeography* **4**, 275–294.

Veblen, T.T., Kitzberger, T. and Lara, A. 1992: Disturbance and forest dynamics along a transect from Andean rain forest to Patagonian shurbland. *Journal of Vegetation Science* **3**, 507–520.

Veblen, T.T. and Lorenz, D.C. 1987: Post-fire stand development of *Austrocedrus–Nothofagus* forests in Patagonia. *Vegetatio* **78**, 113–126.

Veblen, T.T. and Lorenz, D.C. 1988: Recent vegetation changes along the forest/steppe ecotone in northern Patagonia. *Annals of the Association of American Geographers* **78**, 93–111.

Veblen, T.T. and Markgraf, V. 1988: Steppe expansion in Patagonia? *Quaternary Research* **30**, 331–338.

Veblen, T.T. and Stewart, G.H. 1982: On the conifer regeneration gap in New Zealand: the dynamics of *Libocedrus bidwillii* stands on South Island. *Journal of Ecology* **70**, 413–436.

Verghese, B.G. and Ramaswamy, I. 1993: *Harnessing the eastern rivers. Regional cooperation in South Asia.* Konark Publishers, Kathmandu.

Viazzo, P.P. 1989: *Upland communities: environment, population and social structure in the Alps since the sixteenth century.* Cambridge University Press, Cambridge.

Villalba, R. 1995: Climatic influences on forest dynamics along the forest–steppe ecotone in northern Patagonia. PhD dissertation, Department of Geography, University of Colorado at Boulder.

Villalba, R. and Veblen T.T. 1997a: Regional patterns of tree population age structure in northern Patagonia: climatic and disturbance influences. *Journal of Ecology* **85**, 113–124.

Villalba, R. and Veblen, T.T. 1997b: Spatial and temporal variation in *Austrocedrus* growth along the forest–steppe ecotone in northern Patagonia. *Canadian Journal of Forest Research* **27**, 580–597.

Villalba, R. and Veblen, T.T. 1998: Annual versus decadal scale climatic influences on tree establishment and mortality in Northern Patagonia. In Beniston, M. and Innes, J.I. (eds), *Impacts of past, present and future climatic variability and extremes on forests.* Springer-Verlag, Heidelberg and New York, 145–170.

Vinnikov, K.Ya., Robock, A., Stouffer, R.A. and Manabe, S. 1996: Vertical patterns of free and forced climate variations. *Geophysical Research Letters* **23**, 1801–1804.

von Storch, H., Zorita, E. and Cubasch, U. 1991: *Downscaling of global climate change estimates to regional scales: an application to Iberian rainfall in wintertime,* Max-Planck Institut für Meteorologie Report 64, Hamburg.

Vonder Mühll, D. and Holub, P. 1995: Borehole logging in Alpine permafrost, Upper Engadine, Swiss Alps. *Permafrost and Periglacial Processes* **3**, 125–132.

Vonder Mühll, D., Hoelzle, M. and Wagner, S. 1994: Permafrost in den Alpen. *Die Geowissenschaften* **12**, 149–153.

Wade, L.K. and McVean, D.N. 1969: *Mt Wilhelm studies: the alpine and subalpine vegetation.* Australian National University, Canberra.

Wagenbach, D. 1989: Environmental records in alpine glaciers. In Oeschger, H. and Langway, C.C., Jr (eds), *The environmental record in glaciers and ice sheets.* Wiley, Chichester, 69–83.

Walsh, K. and McGregor, J.L. 1995: January and July climate simulations over the Australian region using a limited area model. *Journal of Climate* **8**, 232–247.

Walter, M. and Broome, L. 1998: Snow as a factor in animal hibernation and dormancy. In Greene, K. (ed.), *Snow: a natural history; an uncertain future.* Australian Alps Liaison Committee, Canberra, 165–191.

Wang, Z. 1993: The glacier variation and influence since little ice age and future trends in northwest region, China. *Scientia Geographica Sinica* **13**, 97–104.

Wardle, P. 1973: New Zealand timberlines. *Arctic and Alpine Research* **5**, 127–136.

Wardle, P. 1974: Alpine timberlines. In Ives, J.D. and Barry, R.G. (eds), *Arctic and alpine environments.* Methuen, London, 372–402.

Warren, H.E. and S.K. LeDuc 1981: Impact of climate on energy sector in economic analysis, *Journal of Applied Meteorology* **20**, 1431–1439.

WCMC (World Conservation Monitoring Centre) 1992: *The WCMC biodiversity map library: availability and distribution of GIS datasets.* WCMC, Cambridge.

Webb, T. 1987: The appearance and disappearance of major vegetation assemblages long-term vegetational dynamics in eastern North America. *Vegetation* **69**, 177–187.

Webb, T., Bartlein, P.J., Harrison, S.P. and Anderson, K.H. 1993: Vegetation, lake levels and climate in eastern North America for the last 18,000 years. In Wright, H.E., Kutzbach, J.E., Webb III, T., Ruddiman, W.F., Street-Perrott, F.A. and Bartlein, P.J. (eds), *Global climates since the last glacial maximum.* University of Minnesota Press, Minneapolis, 415–467.

Weber, R., Talkner, P. and Stefanicki, G. 1994: Asymmetric diurnal temperature change in the Alpine Region. *Geophysical Research Letters* **21**, 673–676.

Webster, P.J. and Streten, N.A. 1978: Late Quaternary ice age climates of tropical Australia: interpretations and reconstructions. *Quaternary Research* **10**, 279–309.

Weihe, W.H. 1979: Climate, health and disease. In *World Climate Technical Conference.* World Meteorological Organization, Geneva, 313–368. (Overview paper 13, WMO No. 537.)

WGMS 1993: Glacier Mass Balance, Bulletin 2, ed. W. Haeberli, E. Herren and M. Hoelzle. World Glacier Monitoring Service, ETH Zürich.

Whetton, P.H., Haylock, M.R. and Galloway, R. 1996: Climate change and snow cover duration in the Australian Alps. *Climatic Change* **32**, 447–479.

Whittemore, A.S. 1985: Air pollution and respiratory disease. *Annual Review of Public Health* **2**, 397–429.

WHO 1990: *Potential health effects of climatic change.* Report of a WHO Task Group. World Health Organization, Geneva.

Wigley, T.M.L., Jones, P.D., Briffa, K.R. and Smith, G. 1990: Obtaining sub-grid scale information from coarse-resolution general circulation model output. *Journal of Geophysical Research* 1943–1953.

Wigley, T.M.L., Jones, P.D. and Kelly, P.M. 1980: Scenario for a warm, high CO_2-world, *Nature* **283**, 17–21.

Wigley, T.M.L. and Kelly, P.M. 1990: Holocene climatic change, ^{14}C wiggles and variations in solar irradiance. *Philosophical Transactions of the Royal Society, London* **A330**, 547–560.

Wilbanks, T.J. 1992: Energy policy responses. Concerns about global climate change. In Majundar, S.K., Kalkstein, L.S., Yarnal, B.M., Miller, E.W. and Rosenfeld, L.M. (eds), *Global climate change: implications, challenges and mitigation measures.* Pennsylvania Academy of Science Press, Easton, PA, 453–470.

Wilks, D. S. 1989: Statistical specification of local surface weather elements from large-scale information. *Theoretical and Applied Climatology* **38**, 119–134.

Williams, J. 1980: Anomalies in temperature and rainfall during warm arctic seasons as a guide to the formulation of climate scenarios. *Climatic Change* **19**, 249–266.

Williams, R.J. 1990: Growth of subalpine snowgrass and shrubs following a rare occurrence of frost and drought in south-eastern Australia. *Arctic and Alpine Research* **22**, 412–422.

Williams, R.J. and A.B. Costin 1994: Alpine and subalpine vegetation. In Groves, R.H. (ed.), *Australian vegetation*, 2nd edition. Cambridge University Press, Melbourne, 467–500.

Wilson, C.A. and Mitchell, J.F.B. 1987: Simulated climate and CO_2-induced climate change over Western Europe, *Climatic Change* **10**, 11–42.

Witmer, U., Filliger, P., Kunz, S. and Kung, P. 1986: *Erfassung, Bearbeitung und Kartierung von Schneedaten in der Schweiz.* Geographica Bernensia G25. University of Bern, Bern.

WMO/UNESCO/UNEP/ICSU 1995: GCOS/GTOS plan for terrestrial climate-related observations; Version 1.0. GCOS 21, WMO/TD – No. 721, UNEP/EAP.TR/95-07.

Wolter K. and Timlin, M.S. 1998: Measuring the strength of ENSO events: how does 1997/98 rank? *Weather* **53**, 47–52.

Wood, F.B. 1990: Monitoring global climate change: the case of greenhouse warming. *Bulletin of the American Meteorological Society* **71**, 42–52.

Woodward, F.I. 1987: *Climate and plant distribution.* Cambridge University Press, Cambridge.

Woodward, F.I., Smith, T.M. and Emanuel, W.R. 1995: A global primary productivity and phytogeography model. *Global Biogeochemical Cycles* **9**, 471–490.

World Bank 1995: *Towards environmentally sustainable development in sub-Saharan Africa.* World Bank, Washington, DC.

World Food Institute 1988: *World food trade and U.S. agriculture 1960–1987.* Iowa State University, Ames.

WRI (World Resources Institute) 1996: *World resources 1996-1997.* Oxford University Press, Oxford.

Yoshino M., Jilan, S. and Lee, B.L. 1998: Temperate Asia. In Watson, R.T., Zinyowera, M. and Moss, R. (eds), *The regional impacts of climate change.* Cambridge University Press, Cambridge, 355–379.

Yoshino, M., Makita, H., Kai, K., Kobayashi, M. and Ono, Y. 1980: Bibliography on mountain geoecology in Japan. *Climatological Notes Tsukuba* **25**, 1–111.

Yoshino, M., Horie, T., Seino, H., Tsujii, H., Uchijima, T. and Uchijima, Z. 1988: The effects of climatic variations on agriculture in Japan. In Parry, M.L., Carter, T.R. and Konijn, N.T. (eds), *In the impact of climatic variations on agriculture*, vol. 1, *Assessments in cool temperature and cold regions.* Kluwer, Dordrecht, 723–868.

Zinyowera, M.C., Jallow, B.P., Maya, R.S. and Okoth-Ogendo, H.W.O. 1998: Africa. In *The Regional Impacts of Climate Change.* Cambridge, UK, pp. 29–84.

Zumbühl, H.J. 1988: Der Rhonegletscher in den historischen Quellen. *Die Alpen* **64**, 186–233.

Zumbühl, H.J. and Holzhauser, H. 1988: Alpengletscher in der Kleinen Eiszeit. *Die Alpen* **64**, 322.

Index